Whitehall Paper 99

Necessary Heresies

Challenging the Narratives Distorting Contemporary UK Defence

Justin Bronk and Jack Watling

www.rusi.org

Royal United Services Institute for Defence and Security Studies

Necessary Heresies: Challenging the Narratives Distorting Contemporary UK Defence
First published 2021

Whitehall Papers series

Series Editor: Professor Malcolm Chalmers
Editor: Dr Emma De Angelis

RUSI is a Registered Charity (No. 210639)
Paperback ISBN [978-1-032-26667-1] eBook ISBN [978-1-003-28934-0]
Published on behalf of the Royal United Services Institute for Defence and Security
Studies
by
Routledge Journals, an imprint of Taylor & Francis, 4 Park Square, Milton Park, Abingdon
OX14 4RN

Image courtesy of Kenjo/Adobe Stock

SUBSCRIPTIONS
Please send subscription order to:

USA/Canada: Taylor & Francis Inc., Journals Department, 325 Chestnut Street, 8th Floor,
Philadelphia, PA 19106 USA

UK/Rest of World: Routledge Journals, T&F Customer Services, T&F Informa UK Ltd,
Sheepen Place, Colchester, Essex, C03 0LP UK

Contents

About the Authors

Justin Bronk is the Research Fellow for Airpower and Technology in the Military Sciences team at RUSI. He is also Editor of the *RUSI Defence Systems* online journal. Justin's particular areas of expertise include the modern combat air environment, Russian and Chinese ground-based air defences and fast jet capabilities, unmanned combat aerial vehicles and novel weapons technology.

Justin is a part-time doctoral candidate at the Defence Studies Department of King's College London. He holds an MSc in the History of International Relations from the London School of Economics and a BA (Hons) in History from York University. Justin is also a private glider and light aircraft pilot.

Sidharth Kaushal is the Research Fellow for Sea Power in the Military Sciences team at RUSI. Sidharth's research covers the impact of technology on maritime doctrine in the 21st century and the role of sea power in a state's grand strategy. Sidharth holds a doctorate in International Relations from the London School of Economics, where his research examined the ways in which strategic culture shapes the contours of a state's grand strategy.

Nick Reynolds is the Research Analyst for Land Warfare at RUSI. His research interests include land power, wargaming and simulation. Prior to joining RUSI, he worked for Constellis. Nick holds a BA in War Studies and an MA in Conflict, Security and Development from King's College London. During his studies, he was Head of Operations of the KCL Crisis Team, which organises large-scale crisis simulation events.

Peter Roberts was the Director of Military Sciences at RUSI until December 2021, having been the Senior Research Fellow for Sea Power and C4ISR since 2014. He has researched and published on a range of subjects, including strategy and philosophy, contemporary war, military doctrine and thinking, command and control, naval warfare, ISR, professional military education and disruptive warfare techniques. He lectures, speaks and writes on these topics as well as regularly providing advice for both UK and foreign governments.

Previously, Peter was a career Warfare Officer in the Royal Navy, serving as both a Commanding Officer, National Military Representative and in a variety of roles with all three branches of the British armed forces, the US Coast Guard, US Navy, US Marine Corps and intelligence services from a variety of other states. He served as chairman for several NATO working groups and Five Eyes maritime tactics symposia. His military career included service in Hong Kong, the Baltics, Kenya, the Former Republic of Yugoslavia, Iraq, South Africa, Pakistan and Oman, interspersed with deployments in the GIUK gap and the Persian Gulf.

Peter has a Master's in Defence Studies and a PhD in Politics and Modern History. He is a Visiting Professor of Modern War at the French Military Academy.

Alexandra Stickings is Space Strategy Lead at Frazer-Nash Consultancy, working as part of the strategic advisory team across the space, defence and government sectors. Prior to this, she was Research Fellow for Space Policy and Security in the Military Sciences team at RUSI, where she remains an Associate Fellow. Alexandra's expertise covers military space programmes, space warfare, counterspace capabilities, space situational awareness, arms control, and the intersection of space and missile defence. She has written for a range of publications and is a frequent speaker at international conferences.

Jack Watling is the Research Fellow for Land Warfare. Jack has recently conducted studies of deterrence against Russia, force modernisation, partner force capacity building, the future of corps operations, the future of fires, and Iranian strategic culture. Jack's PhD examined the evolution of Britain's policy responses to civil war in the early 20th century. Prior to joining RUSI, Jack worked in Iraq, Mali, Rwanda, Brunei and further afield.

INTRODUCTION

JUSTIN BRONK AND JACK WATLING

Humans are naturally drawn to stories, since we use narratives to make sense of a complex world, to order information into chains of causality, and to communicate and respond to ideas.[1] Confronted with novel technologies or tactics, we are often drawn to narrative vignettes of how these capabilities could be employed in order to visualise their effects.[2] However, narratives are not merely descriptive; they implicitly promote frameworks that prompt behaviour and judgement. It has long been recognised in economics that narratives shape expectations, stimulate imagination and guide investment decisions in ways that empirical analysis often struggles to match.[3] Within Defence, the shaping influence of uncritically accepted narratives can have problematic consequences.

In many areas of defence policy, such as cyber warfare, space or novel weapons systems, deep subject matter expertise is required to understand the potential benefits and limitations. The same is true of attempts to assess the policies and actions of strategic competitors with very different cultural and geopolitical viewpoints. Crucial nuances and practical constraints are almost unavoidably lost in translation as senior decision-makers shape policy and generalists rewrite doctrine and strategy documents based on their own understanding of briefings given by specialist practitioners and subject matter experts. This tendency is exacerbated by a natural inclination to over-hype the potential for novel technologies or strategies to provide transformative effects. Incompatible

---

[1] Jonathan Gottschall, *The Storytelling Animal: How Stories Make Us Human* (New York, NY: Harcourt, 2013).
[2] A phenomenon that explains the success of books like P W Singer and August Cole, *Ghost Fleet: A Novel of the Next War* (New York, NY: Houghton Mifflin, 2015).
[3] Jens Beckert and Richard Bronk (eds), *Uncertain Futures: Imaginaries, Narratives, and Calculation in the Economy* (Oxford: Oxford University Press, 2018).

political demands for increased financial and manpower efficiency, equipment modernisation, improved force readiness, resilience and constant engagement in global competition incentivise defence planners and policymakers to seek silver bullet solutions. Once policy has been stated on an issue, further nuances and important caveats are often lost as the wider policy community try to tailor their own outputs to align with what they perceive as the new high-level consensus. As such, the narratives that end up shaping much of the 'coal face' work in Defence are not the (usually) nuanced and well-caveated statements on novel technologies, domain activities or adversary tactics prepared and published by specialists. Instead, they are often mantras or collective 'received wisdom' that in practice have been oversimplified or distorted by repeated translation, repetition and transmission.

In the Chief of the Defence Staff's annual lecture, delivered at RUSI in December 2020, General Sir Nick Carter asserted that 'our rivals seek to win without resorting to war'. He suggested that there is 'a clear trend towards military action that uses the cognitive elements of war with arms-length instruments like drones and mercenaries to provide a plausible degree of deniability and strategic ambiguity – thus enabling intervention without the risk of entanglement'. He argued that 'competition below the threshold of war is not only necessary to deter war, it is also necessary to prevent one's adversaries from achieving their objectives in fait accompli strategies as we have seen in Crimea, Ukraine, Libya and the South China Sea', and posited that 'the means to control others – principally through the application of technology - is the crux of the matter'.[4]

The authors contend that narratives such as these are not only fundamentally unsound but also produce potentially harmful distortionary effects throughout Defence. Far from seeking victory without war, Russia used overt conventional military force against Georgia and Ukraine to seize territory, as did China in Ladakh,[5] and in the South China Sea to militarily occupy atolls. Russian operations in Syria,[6] Libya[7] or the Donbas have not evaded 'entanglement'.[8] The insinuation that the bombardment at Zelenopillya[9] or the vicious battle

---

[4]   Ministry of Defence (MoD) and Nick Carter, 'Chief of Defence Staff Speech RUSI Annual Lecture', 17 December 2020, <https://www.gov.uk/government/speeches/chief-of-defence-staff-at-rusi-annual-lecture>, accessed 10 July 2021.
[5]   Shashank Joshi, 'A Border Dispute Between India and China is Getting More Serious', *The Economist*, 30 May 2020.
[6]   *BBC News*, 'Turkey Shoots Down Russian Warplane on Syria Border', 24 November 2015; Thomas Gibbons-Neff, 'How a 4-Hour Battle Between Russian Mercenaries and U.S. Commandos Unfolded in Syria', *New York Times*, 24 May 2018.
[7]   US Africa Command Public Affairs, 'Russia and the Wagner Group Continue to Be Involved in Ground, Air Operations in Libya', 24 July 2020, <https://www.africom.

for Donetsk airport[10] were part of fait accompli strategies or were actions which fell short of warfighting is indefensible. While adversary campaigns have used novel technologies including uncrewed aerial vehicles and cyber attacks, it was hard power and not 'the application of technology' that provided the means by which they coerced and controlled their opponents. In order to coerce Kyiv in February 2021, Moscow applied pressure through the build-up of over 100,000 troops on the Ukrainian border,[11] the tangibility of which contrasted starkly with Western statements of solidarity and concern.[12] Similarly, China is able to coerce its neighbours with its maritime militia in the South China Sea,[13] not because it is carefully avoiding escalation, but because of the size of the People's Liberation Army Navy, which makes the Philippines and Vietnam wary of providing a pretext for Beijing to apply overt military force.[14] Russia and Iran have used undeclared military forces in their recent campaigns. But this is not new. The use of undeclared forces featured prominently throughout the Cold War by all sides, from the Korean War and the Vietnam War[15] to the Angolan Civil War[16] or the UK's campaign in Oman.[17]

---

mil/pressrelease/33034/russia-and-the-wagner-group-continue-to-be-in>, accessed 11 July 2021.

[8] Andrew Kramer, 'Fighting Escalates in Eastern Ukraine, Signaling the End to Another Cease-Fire', *New York Times*, last updated 30 April 2021.

[9] Bellingcat Investigative Team, 'Bellingcat Report – Origin of Artillery Attacks on Ukrainian Military Positions in Eastern Ukraine Between 14 July 2014 and 8 August 2014', *Bellingcat*, 17 February 2015.

[10] Amos C Fox, '"Cyborgs at Little Stalingrad": A Brief History of the Battles of Donetsk Airport', Land Warfare Paper No. 125, Institute of Land Warfare, May 2019.

[11] Robin Emmott and Sabine Siebold, 'OFFICIAL Russian Military Build-Up Near Ukraine Numbers More Than 100,000 Troops, EU Says', *Reuters*, 19 April 2021; Cyrus Newlin et al., 'Unpacking the Russian Troop Buildup Along Ukraine's Border', Center for Strategic and International Studies, 22 April 2021.

[12] Andrew Hilliar, 'Biden Offers Ukraine "Unwavering Support" Over Russia Standoff', *France24*, 3 April 2021.

[13] *BBC News*, 'South China Sea Dispute: Huge Chinese "Fishing Fleet" Alarms Philippines', 21 March 2021.

[14] Sidharth Kaushal and Magdalena Markiewicz, 'Crossing the River by Feeling the Stones: The Trajectory of China's Maritime Transformation', *RUSI Occasional Papers* (October 2019).

[15] Francis X Clines, 'Russians Acknowledge a Combat Role in Vietnam', *New York Times*, 14 April 1989.

[16] Central Intelligence Agency, Directorate of Intelligence, 'Intelligence Memorandum: Soviet and Cuban Intervention in the Angolan Civil War', March 1977, <https://www.cia.gov/readingroom/docs/DOC_0000518406.pdf>, accessed 11 July 2021.

[17] John Akehurst, *We Won a War: The Campaign in Oman, 1965–1975* (Salisbury: Michael Russell, 1982).

Through years of repetition, narratives about the rapidly changing character of warfare and the transformative effects of novel technologies have become akin to gospel truths, enshrined in policy documents. The Integrated Operating Concept, published in September 2020, stated that 'old distinctions between "peace" and "war" ... are increasingly out of date'.[18] Just as General Carter argued that technology shall provide the means of control, so too does the Integrated Operating Concept 'require us to embrace combinations of information-centric technologies to achieve the disruptive effect we need'.[19] As General Carter warned, 'our rivals ... developed long range missile systems; they integrated electronic warfare, [and] swarms of drones connected digitally to missile systems and used these to defeat tanks'.[20] And so the Integrated Operating Concept argues that 'expensive, crewed platforms that we cannot replace and can ill afford to lose will be increasingly vulnerable to swarms of self-coordinating smart munitions – perhaps arriving at hypersonic speeds or ballistically from space – designed to swamp defences already weakened by pre-emptive cyber attack'.[21]

The fact that technological change forces militaries to adapt how they fight is axiomatic. Yet, it is striking how a narrative that places a disproportionate emphasis on technological transformation betrays a poor understanding of how these technologies work, and consequently misrepresents their likely effects. To take the example from the Integrated Operating Concept above, while swarming munitions, smart munitions, hypersonic and ballistic missiles, and cyber attacks will feature on the future battlefield, many of these properties are mutually exclusive. Self-coordinating smart munitions – let alone hypersonic ones – cannot be cheap enough for most states to field them in sufficient numbers to 'swarm' the battlefield. This is especially true given that, in Defence, capability ambitions consistently outstrip available funding.[22] Nevertheless, such capabilities are regularly combined in a form of word soup as the basis for transformative narratives in speeches by senior officials and high-level policy documents.

Although the misconceptions that this Whitehall Paper seeks to confront have been widely promulgated by senior officials publicly, it is important to recognise that many of those who propound them take a

---

[18] MoD, 'Integrated Operating Concept', August 2021, p. 5.
[19] *Ibid.*, p. 16.
[20] MoD and Carter, 'Chief of Defence Staff Speech RUSI Annual Lecture'.
[21] MoD, 'Integrated Operating Concept', p. 6.
[22] Claire Mills, Louisa Brooke-Holland and Nigel Walker, 'A Brief Guide to Previous British Defence Reviews', Briefing Paper No. 07313, House of Commons Library, 26 February 2020.

more nuanced view in private. Nevertheless, that nuance is often absent from policy documents because of biases that perennially afflict bureaucracies. The result for Defence is that once adopted by decision-makers, phrases like 'competition below the threshold of war' gain a gravitational power that distorts surrounding debate. It can even lead to the recycling of outright disinformation – such as the existence of a Gerasimov Doctrine – which survives in the discourse among non-specialist officials years after the original error was corrected.[23] In bureaucracies, once ideas are codified into policy documents, those who challenge the consensus risk being bypassed as obstructionists or side-lined as heretics. However, it is the contention of the authors that challenging some of the narratives which form part of the current consensus view in UK defence discourse is a necessary heresy.

The authors do not dispute that the character of warfare is evolving. Emerging technologies are already changing how militaries are structured and how they will fight in the future. However, many of the interpretations of how emerging technologies and supposedly novel adversary activities will shape the future defence and security environment in the current discourse are provably false. This paper is an attempt to correct some of these misleading narratives before they drive acquisition decisions that undermine the UK's conventional forces. The related arguments it seeks to refute are:

- That new domains of warfare, from space to cyberspace and 'information' render traditional land, air and maritime platforms as 'sunset' capabilities.
- That adversaries prefer to fight in the 'grey zone', rather than being forced to pursue inefficient strategies because of the effective application of conventional deterrence.
- That technology will eliminate the relevance of critical combat mass in future operations.
- That information operations, cyber attacks and precision-strike capabilities can in themselves decisively cripple a state's capacity and will to fight.

Confronting these propositions is made difficult by the fact that they arise from a constellation of conceptual and technical misconceptions. The mischaracterisation of the grey zone, for instance, is both a product of conceptual inaccuracy in how adversaries approach escalation and of an

---

[23] Mark Galeotti, 'I'm Sorry for Creating the "Gerasimov Doctrine"', *Foreign Policy*, 5 March 2018.

overestimation of the efficiency and efficacy of grey-zone tools such as cyber operations. The belief that mass is no longer relevant to operations arises from inflated expectations as to what can be delivered by technology at an affordable cost, and a failure to map out the tasks that fall to militaries that will continue to require people to carry them out. Furthermore, some underlying errors inform more than one of the conclusions that this book seeks to refute. For this reason, the chapters do not directly rebut the conclusions above but are instead aimed at correcting the underlying technical and conceptual misconceptions.

In Chapter I, 'The Slow and Imprecise Art of Cyber Warfare', Justin Bronk and Jack Watling seek to unpack how cyber attacks work and what they require to succeed, in order to provide a measured assessment of what cyber capabilities can actually deliver in a military context. Few subjects better highlight the gulf separating the technical and policy communities. The authors contend that, against hardened military systems, cyber attacks take far longer to prepare than political planning for conflict usually provides and, once emplaced, enduring capability is difficult to assure. The result is that while cyber warfare can provide unexpected opportunities when used in combination with conventional military capabilities, it is slow, unpredictable and demands exceptionally long planning cycles to produce results. It is a potentially potent enabler and will continue to form an important component of the modern security and defence toolbox, but it is not interchangeable with, nor a replacement for, kinetic power. Moreover, if defence establishments do not establish an operational echelon that can link cyber technical experts into the planning cycles for other domains, then cyber capabilities are unlikely to be in place when and where they are needed.

The long lead times in emplacing cyber capabilities mean that such operations must be carried out long before the outbreak of hostilities. Indeed, a cyber attack is often the main example desired to describe grey-zone activity, prosecuted with hostile intent but below the threshold of warfighting. The discourse surrounding the grey zone has become amorphous but has a tendency to frame certain capabilities as grey-zone 'tools' or 'strategies'. The problem with this framing, as Sidharth Kaushal explores in Chapter II, 'The Grey Zone Is Defined by the Defender', is that the threshold between warfighting and the grey zone is not fixed. A cyber attack could, in fact, prompt a conventional military response. What determines whether an attack is 'below the threshold' of armed conflict is whether the defender is prepared to escalate to violence in retaliation. States do not pursue 'grey-zone strategies' because they are necessarily preferable, but rather because they have judged that certain activities are what can be pursued without unacceptable consequences. To this end, the grey zone can in fact include kinetic fighting if the defender is deterred from retaliating with conventional forces. Grey-zone activities are often an inefficient way of prosecuting a campaign and are a function of the relative balance of conventional deterrence between parties. The policy question therefore is not how to deter grey-zone activity, but which activities will elicit a response, and which will not.

That conventional hard power underpins deterrence and buys the capacity to constrain the parameters of the grey zone means that the credibility of the UK's conventional forces bears examination. There is a pervasive trend in UK defence discourse to assert that technology can deliver more capability with fewer personnel. In Chapter III, 'Doing Less with Less in the Land Domain', Nick Reynolds explores the practical limits to reductions in conventional forces before capability suffers from exponential decay. While acknowledging that technology can allow fewer personnel to deliver greater effects on the battlefield, there are irreducible minimums to be able to project power in the land domain. From a force generation point of view there is a requirement to train, project, sustain and deploy a force, which must then have a presence across the ground. A deployable division is the minimum viable force for participation in even limited wars. However, if a force lacks further reserves, its tactical options can be severely constrained because preserving the force becomes a pre-eminent task. Historical campaigns are replete with early setbacks, and this is even more likely in the future, given the lack of data to guide expectations about how technologies will manifest in combat. In this context, the capacity to reconstitute a first echelon and being large enough to absorb failures is vital to having credible military forces and therefore deterring adversaries into using grey-zone tools.

The scale at which transformative technologies are anticipated to populate the future battlefield is vastly inflated, overlooking their cost and complexity. In Chapter IV, 'Swarming Munitions, UAVs and the Myth of Cheap Mass', Justin Bronk examines the promised revolution in terms of increased combat mass through the use of swarming munitions and so-called 'attritable, reusable unmanned aerial vehicles' (UAVs).[24] There are important grains of truth fuelling the narrative. Prototypes for attritable, reusable UAVs and swarming munitions are being rapidly developed by several states and offer novel tactical options and potential efficiency gains across a range of key mission sets. These weapon classes will significantly alter the way in which air forces conduct operations at the tactical and even operational levels. However, despite being

---

[24] For example, RAF Chief of the Air Staff, ACM Mike Wigston quoted in Aaron Mehta, 'Britain's Royal Air Force Chief Talks F-35 Tally and Divesting Equipment', *Defense News*, 10 May 2021.

technologically feasible, many of the visions for how such capabilities might be employed at scale greatly overstate the cost and performance efficiencies achievable. The result is that while militaries will gain significant flexibility from these novel weapons systems, they will not be applicable to all mission sets and are unlikely to be affordable in the quantities envisaged. Thus, they should not be assumed to offer a large-scale replacement for traditional combat aircraft, nor do these technologies represent a silver bullet solution to a persistent lack of combat air mass. Finally, if states will have limited arsenals of these capabilities, then they will need to be very discerning as to what targets they employ them against.

Exaggerated expectations as to the volume and effects deliverable by novel strike capabilities are arguably leading to dangerous assumptions about mutual deterrence. The threats posed by long-range precision strike – combined with cyber attacks against critical national infrastructure – are often presented as capabilities that could win a war before it starts by preventing an adversary from being able to deploy. This is tied to the notion that the infliction of major casualties would politically cripple an adversary's will to fight. In Chapter V, 'The Lights May Go Out, But the Band Plays On', Peter Roberts argues that, throughout history, states have proven highly robust and adapt quickly in response to shocks. Unless physically occupied, they are usually able to reconstitute and conduct prolonged operations. Furthermore, casualty tolerance is not a fixed variable but is highly context dependent. In any major conflict, casualty tolerance often increases drastically. Even in small conflicts, states are able to sustain a high level of attrition and maintain support for operations when the public has confidence in the cause for which they are fighting. The implications are that, when evaluating the likely outcome and risk of conflict, policymakers should pay careful attention to the relative strengths of their second echelons – they should not just ask who will win the first battle, but extend their analysis to the second. They should also work hard to ensure that there is public support for key causes, rather than self-deter their policy by assuming a blanket intolerance of casualties.

In contrast to the resilience of states, space-based infrastructure is robust in its component parts but systemically fragile. Although redundancy in a growing number of constellations is increasing the challenge of countering specific military space-based capabilities, the risk of denying access to orbits through the creation of debris creates a form of mutual deterrence. Dependence on and increasing capabilities in space are seeing its progressive militarisation and fuelling a misplaced fear of kinetic anti-satellite capabilities and weaponised vehicles in the space domain.[25] In Chapter VI, 'In Space, No One Will See You Fight',

Alexandra Stickings argues that the fragility of access to the space domain creates strong disincentives for states to conduct kinetic operations in orbit. While these may occur on a highly limited basis, most conflict in space will comprise manoeuvring for advantageous orbits and the application of non-kinetic effects to enable operations in the other domains. The process of competition will occur continuously, while targeting in space will be strongly tied to the imperatives created by terrestrial operations. Thus, the key question for understanding warfare in space is less the interaction between orbital systems, but rather the effects these systems have on the ground. Indeed, the highest-intensity conflict for access to space may well comprise the targeting and occupation of the ground-based infrastructure. Therefore, space should not be a siloed specialism but must be integrated into other warfighting domains. Most importantly, space may stand as an important example of how political constraints will remain a feature of military decision-making even in large-scale warfare, a fact often ignored in military concepts and wargames.

While the wholesale destruction of space-based infrastructure is unlikely, denial of access for a limited duration at critical moments from defined geographic areas pose a serious threat to the assurance of communications on the battlefield. There is a growing obsession across militaries with connecting all parts of the force into an 'any-sensor-to-any-shooter' network. While ensuring that systems can talk to one another when links are available, the implications of this network are highly uneven across the domains. In Chapter VII, 'More Sensors Than Sense', Jack Watling explores the challenges in sharing data around a battlefield within a contested electromagnetic spectrum and argues that the advantages gained from interconnectedness are highly context dependent. While transformative in the maritime domain, there are operational limitations on its usefulness to air forces, and massive technical hurdles to expecting reliable advantage from such systems in land operations. This is not to say that a networked force will not have advantages, but that those advantages will be more incremental, contextually dependent and less assured than is widely supposed. Thus, the regular use of 'combat cloud'-enabled concepts in narrative descriptions of future military operations needs careful scrutiny if dangerous distortions to force planning are to be avoided.

As the UK military seeks to modernise in accordance with the Integrated Review, it is vital that this is done with a sound conceptual

---

[25] For example, *BBC News*, 'UK and US Say Russia Fired a Satellite Weapon in Space', 23 July 2020; Kim Sengupta, 'UK Seeks to Prevent Space Arms Race After Russia Launches Anti-Satellite Missiles', *The Independent*, 26 August 2020.

understanding of both how adversaries are operating and what is technologically possible. If the UK is to increasingly deploy forces to deter Russian and other hostile activity and to compete without escalation into warfighting, it is vital that military planners understand what the adversary is doing and why. Moreover, while investment in novel technologies and force modernisation is key, policy must be tempered by contextual and technical understanding. Policy based on uncritically repeated and amplified narratives risks overinvestment in capabilities with insufficient deterrent value in the eyes of adversaries, and would fail to deliver victory unless fielded in coordination with more traditional tools of hard power.

I. THE SLOW AND IMPRECISE ART OF CYBER WARFARE

JUSTIN BRONK AND JACK WATLING

Sally Walker, former Director of Cyber at GCHQ, has stated that cyber attacks 'can have impact in the real world and you can do it at scale'.[1] Because cyber attacks can target everything from financial systems and critical national infrastructure to political leaders and legal institutions – undermining trust and the rule of law – they can have an 'attritional' effect on the cohesion of states to which open societies 'are uniquely vulnerable'.[2] Such warnings over the years, combined with high-profile incidents like the WannaCry ransomware attack against the NHS, have meant that military leaders have recognised the importance of the 'cyber domain'.[3] When new forms of conflict first emerge, however, there is almost always a period of inflated expectation.[4] For the small community within Defence who have worked in the margins to explore the novel capability, there is a tendency to hype its effects and downplay its limitations in order to gain the attention of the wider defence and security community and secure resources within the bureaucracy. For that wider community – lacking an understanding of the capability – it is often much easier to visualise the potential threats posed by novel weapons than the challenges involved in employing them effectively. Cyber warfare today is arguably at the peak of this inflated discourse,

---

[1] Sally Walker, 'Into the Grey Zone Podcast: Episode Five – Cyber Power (Part II)', *Sky News*, 3 June 2021, 06:00, <https://news.sky.com/story/into-the-grey-zone-podcast-episode-five-cyber-power-part-ii-12212228>, accessed 26 July 2021.
[2] *Ibid.*, 10:30.
[3] National Audit Office, *Investigation: WannaCry Cyber Attack and the NHS*, HC 414 (London: National Audit Office, 2018).
[4] This is visualised in Gartner, 'Gartner Hype Cycle', <https://www.gartner.com/en/research/methodologies/gartner-hype-cycle>, accessed 30 March 2021.

where its importance is recognised but its implications remain poorly understood beyond the limited community of technical specialists.

In briefings, military commanders often characterise cyber capabilities by the speed, scale and sophistication of the effect. Outlining the challenge of 'convergence' between warfighting domains, Lieutenant General Eric Wesley described how 'cyber is moving at milliseconds, ships are moving at 30 knots, vehicles at 30 km. The ability to synchronise that might be a bit of a stretch'.[5] As well as an expectation of near-instantaneous effects, there is a persistent fear of the devastation that could be wrought. General Wesley Clark, former NATO Supreme Commander Europe, warned that 'it's possible to shut down critical infrastructure, refineries, businesses, banks, transportation – but most importantly, the electricity grid. It's been done by the Russians in the Ukraine'.[6] Finally, there is a great deal of anxiety in militaries about the sophistication of cyber effects against conventional military systems. General John Allen, for instance, hypothesised about an incident in which a US Navy vessel had its defensive systems rendered ineffective by cyber attack, making it vulnerable to conventional munitions.[7]

A more thorough examination of the mechanics of cyber warfare renders much of this military discourse highly hyperbolic. Although data can travel quickly, cyber attacks are slow to develop and deliver, and often unpredictable and isolated in their effects. The challenge with integrating cyber effects into multi-domain operations is the opposite of the problem suggested by Wesley – that cyber attacks take far too long to easily synchronise with conventional operations. As Walker noted, 'it just doesn't relate to conventional warfare'.[8] A realistic appreciation of the impact of cyber capabilities on the future of warfare must begin with a concrete understanding of how cyber attacks work. Appreciating what is required to attack a digital system reveals its limitations and so allows for the potential impact of such techniques on warfare to be put into perspective. This chapter seeks to explain how cyber attacks function and suggests their likely trajectory on the future battlefield.[9]

---

[5] US Department of Defense, 'AUSA Global Force Symposium: Day 3 – Opening Remarks and Keynote Speaker', 18 March 2021, 35:00, <https://dod.defense.gov/News/Special-Reports/Videos/?videoid=668339>, accessed 26 July 2021.

[6] Mandi Kogosowski, '"We're on the Brink of an Enormous Cyber Catastrophe"', *Israel Defense*, 4 June 2021, <https://www.israeldefense.co.il/en/node/49194>, accessed 26 July 2021.

[7] John R Allen and Amir Hussain, 'On Hyperwar', *Proceedings Magazine* (Vol. 143, No. 7, 2017), p. 1373.

[8] Walker, 'Into the Grey Zone Podcast', 13:49.

[9] Cyber practitioners are necessarily very precise in the language they employ. This chapter is intended to help non-cyber practitioners understand key concepts and

The Anatomy of a Cyber Attack

Cyber activity is primarily concerned with the virtual dimension,[10] which is characterised by its dependence on invented languages.[11] Information encoded in these invented languages comprises data.[12] Military cyber attacks seek to achieve at least one of three things: to access data that an adversary does not wish to share; to change the data that an adversary holds; or to destroy an adversary's data.[13] The first requirement for conducting a cyber attack is to identify that an adversary holds data. For example, a range table for a munition in a printed handbook exists in the physical dimension as paper and in the cognitive dimension if it has been memorised by an operator, but not in the virtual dimension. If, however, the range table has been translated into data and is stored on a digital device, it exists in the virtual dimension and can be subjected to cyber attack. If this data is accessible as part of a fire control system supporting geographically dispersed artillery pieces, then that data must be remotely accessible via a network. It can therefore be attacked remotely.

Since the virtual dimension is fabricated, it is usually not immediately apparent to an external party what systems have been digitised, what they are connected to, or what languages are used to store the data and manage the network. Furthermore, programming languages do not simply store information but also contain grammar for describing commands that the system can execute. These commands describe how the data is to be stored, who has permission to access it, and what checks are carried out to determine who each user is. All of these facets of a network are invisible to an intruder until they have interacted with part of that

---

uses analogy and imprecise terms to aid accessibility. The authors apologise to specialists in the field for the resulting lack of precision.
[10] Ministry of Defence, 'Joint Doctrine Publication 0-30: UK Air and Space Power', 2nd edition, December 2017, p. 17.
[11] David J Peterson, *Art of Language Invention, The: From Horse-Lords to Dark Elves to Sand Worms, the Words Behind World-Building* (London: Penguin, 2015), pp. 21–22.
[12] The subset of invented languages used in computing comprises markup and programming languages. Markup languages allow the programming instructions to not be displayed when displaying data in a non-programming language. See Thomas Powell, *HTML & XHTML: The Complete Reference* (New York, NY: McGraw-Hill, 2003), p. 25. Programming languages use imperative and declarative syntax to describe and command the execution of specified functions, see Michael Gordon, *Programming Language Theory and Its Implementation* (Hoboken: Prentice Hall, 1988).
[13] These offensive aims can be understood as the offensive mirror image of the CIA's triad model of cybersecurity principles: confidentiality; integrity; and availability. See 'What Is Security Analysis?', <https://www.doc.ic.ac.uk/~ajs300/security/CIA.htm>, accessed 21 October 2021.

network. Therefore, having determined that a digital target exists, an attacker must seek to intrude into the network which contains it, and exfiltrate data to determine how that network is constructed and by what languages it is governed.

A lack of information about a network makes designing penetrations that are difficult to detect almost impossible, just as it is very difficult for a non-native speaker of a language to pass as a native speaker.[14] For this reason, the initial penetration and mapping of networks involves an extensive reconnaissance effort. Attackers will often conduct a high volume of unsophisticated attacks on the basis that while only a few will make any headway into the network, they will generate valuable information about which nodes in the network hold administrative privileges and how the protocols that govern the network function.[15] Large-scale unsophisticated attacks produce a constant noise so that adversaries cannot necessarily determine what is being targeted. In many ways, this is because the purpose is to find what targets might be available. This is comparable to a reconnaissance by force, or advance to contact, rather than a deliberate reconnaissance. However, the defenders will try to identify who is perpetrating this reconnaissance in order to anticipate what subsequent, more sophisticated attacks might look like.[16]

Once the shape of a network and the location of data and privileges are identified, an attacker will begin to craft specialised attacks to gain access to those points. This will often require detailed information about the humans who govern these points in the network, and how they interact with it in the physical dimension, because this can provide valuable information about how to attack the system.[17] These deliberate attacks against identified nodes must be carried out without being detected. If an intrusion is detected, the defender can simply sever the node that has been penetrated or change the permissions to deny it any useful access to the rest of the network. At that point, the attacker must

---

[14] Simon Müller, 'Discourse Makers in Native and Non-Native English Discourse', PhD submission to Justus Liebig University Giessen, Amsterdam, 2004; Melissa M Baese-Berk and Tuuli H Morrill, 'Speaking Rate Consistency in Native and Non-Native Speakers of English', *Journal of the Acoustical Society of America* (Vol. 138, No. 3, 2015), pp. 223–28.

[15] H P Sanghvi and M S Dahiya, 'Cyber Reconnaissance: An Alarm Before Cyber Attack', *International Journal of Computer Applications* (Vol. 63, No. 6, 2013), pp. 36–38.

[16] This is the purpose of cyber threat intelligence. See Henry Dalziel, *How to Define and Build an Effective Cyber Threat Intelligence Capability* (Waltham, MA: Syngress, 2015).

[17] Micah Zenko, *Red Team: How to Succeed by Thinking Like the Enemy* (New York, NY: Basic Books, 2015), pp. 171–83.

begin from scratch against a defender that has gained information about the attacker's techniques and signature.

After a critical node is penetrated, data from it must be exfiltrated and analysed to allow the attacker to see further into the network and target the next node up the chain. Ultimately, the aim is for the attacker to gain access to repositories of data, or to components of a system that can control other parts of the system. Whereas most parts of a network have the right to read a limited set of data, some administrative nodes have permission to rewrite the instructions governing the system. Attackers seek access to administrators because these accounts can change or destroy the data or physical infrastructure that constitute the enemy system. For example, an administrator for a fire control system might substitute an accurate range table for an inaccurate one, ensuring the artillery system as a whole misses any target it tries to engage. Alternatively, it could delete the range table so that the system cannot calculate firing solutions. The administrator could even delete the instructions for the system to read a range table, and then destroy the instructions for the system to accept new instructions so that the defender cannot simply upload a new range table when the data is found to be missing.

Almost all components of a network will be subject to varying levels of encryption which must be cracked or bypassed by an attacker at each stage. This could require the entry of a simple password, or for more sensitive and better defended systems, a three-part authentication comprising something known by the individual who owns the node (a password), something they have (a physical key), and something they are (biometric data). Passwords to confirm that someone requesting access to a system is the right individual can be cracked through a brute force (guessing all possible combinations) or analytical attack. The latter method can be invalidated by randomly generating passwords. In practice, any effectively defended system will have millions of possible solutions to passwords at each level, and incorrect entries will attract the attention of defenders. For this reason, attackers will endeavour to bypass passwords through hacking. Hacking involves the interrogation of the code to find ambiguities through which an attack can convince the system that a password is not necessary or has been entered when it has not.[18]

---

[18] There is an alternative distinction drawn between hacking and cracking that is framed in ethical terms, with the former being constructive and the latter destructive. In practice, this ethical distinction – while important in the early days of the internet – has fragmented. See John P Carlin with Garrett M Graff, *Dawn of the Code War: America's Battle Against Russia, China, and the Rising Global Cyber Threat* (New York, NY: Public Affairs, 2018), chapter 1.

To envisage how a hack works, consider the consequences of someone using poor grammar, as in the sentence: 'Correct and verified, the user may decrypt the data'. In conversation, this can be understood to mean 'if the user has been correctly identified and verified, then they may decrypt the data'. However, the sentence could equally be literally interpreted as saying 'the user is correct and verified and may decrypt the data'. A hack would suggest to the system that it should proceed with the latter interpretation. Because complex code involves thousands of lines of instruction, often written by multiple people at different times, there will often be minor contradictions or ambiguities that can be discovered and then leveraged by an attacker.[19]

When these ambiguities and contradictions are part of the code in a system, they are termed 'zero-days' because they are vulnerabilities that were introduced into the system by the way it was designed, and at the time of discovery represent an open vulnerability that has not been patched, leaving defenders with zero days to resolve the problem. By storing zero-day exploits which they discover through reconnaissance efforts, an attacker can subsequently employ them to bypass a system's defences. However, just as the poor grammar cited above can be clarified, so too can a defender clarify their coding to make a system more secure. If, therefore, they identify that a zero-day exploit has enabled penetration of their network, they can often patch the system to edit out and close the vulnerability, or else closely monitor it to gain intelligence on potential ongoing or future attacks. Consequently, attackers must make a trade-off between the value of accessing a system at any given time and the risk of losing the ability to bypass its encryption or being rapidly attributed and potentially counter-attacked if discovered.

Attacking sensitive digital systems is further complicated by the fact that they are often air-gapped, meaning they are not susceptible to remote access. This means that, to attack them, an adversary must either create an artificial connection to the outside world or physically plug into the air-gapped network without being detected. This may require multiple physical penetrations since the first intrusion may simply allow the attacker to collect information about what language the system uses and how it is structured. They must then physically penetrate the system a second time to deliver a designed payload to attack it. If they do not wish to immediately activate the delivered payload, they may need to

---

[19] A common issue is poor hygiene in compound coding for signifier/signified correlation. See Paolo Rocchi, 'The Concepts of Signifier/Signified Revisited', IBM, <http://mcs.open.ac.uk/dtmd/Presentations/Session%202/Rocchi.pdf>, accessed 30 March 2021.

penetrate the system a third time to trigger the attack at the desired point in time.

A cyber attack may therefore be envisaged as comprising seven stages:

1. **Reconnaissance:** to understand the target network.
2. **Weaponisation:** designing a payload to exploit vulnerabilities.
3. **Delivery:** insertion of the payload through the system.
4. **Exploitation:** using new access to identify and target the key node.
5. **Installation:** installing the payload at the target node of the system.
6. **Command and control:** having a means of initiating the payload.
7. **Action on objectives:** having the desired effect on the system.[20]

Attempting to gain entry to a complex defended system may require several iterations of this cycle as the attackers fight their way into the network. Against a complex network with multiple layers, which ultimately ends in an air-gapped component, working into a network may take between one and three years.[21] While embedding malicious capabilities into public utilities or other legacy civilian systems may be more straightforward,[22] attacking military systems is complex, sensitive and has a significant lead time, requiring payloads to be embedded during peacetime if they are to be available during war.

If a cyber attack is prosecuted in peacetime and the payload initiated before the outbreak of war, its effects will either be limited or escalatory. For example, to return to the artillery fire control system, if it were penetrated and the range tables altered prior to a conflict, this could mean that the change would be noticed during an exercise, the range tables corrected and the system security patched prior to a conflict. Alternatively, it could mean that during the same exercise the artillery kills friendly forces. This would lead to an investigation and, if the cause could be attributed to a hostile cyber attack, it would lead to a significant

---

[20] Joseph Raczynski, 'Kill Chain: The 7 Stages of a Cyberattack', *Thomson Reuters*, 12 October 2018.

[21] Author interview with a senior CTI director at a major bank, March 2021; author interview with a senior military cyber warfare officer, February 2021; author interview with a law enforcement officer specialising in counter-cyber operations, January 2021.

[22] House of Commons, 'A Major Cyber Attack on the UK is a Matter of "When, Not If"', Joint Committee on the National Security Strategy, 22 January 2018, <https://houseofcommons.shorthandstories.com/jcnss-cni-report/index.html>, accessed 30 March 2021.

escalation in tensions or could even be interpretated by the victim as a *casus belli*.[23]

Conversely, a cyber payload that sits dormant in a system until activated during a conflict presents other challenges. It must either be activated automatically when certain criteria are met or manually. The latter requires a sustained capacity to connect to the system at short notice to be maintained by the attacker. For military systems – given the changes that occur when a state transitions to war – this may be difficult, requiring the embedding of agents with knowledge of the penetration inside the enemy force. However, a payload that monitors the system so that it can activate under certain conditions is an active rather than passive insertion and is, therefore, more susceptible to detection. It could also activate under circumstances unforeseen by the attacker.

The final limitation of a military cyber capability is that cyber attacks function by inserting data into a system so that it is directed to behave in a specific manner. It is effective not in itself, but because of how the payload interacts with the rest of the system. Because the system is fabricated, however, and the language invented, it is subject to being expanded, divided, updated and changed in fundamental ways. Therefore, a payload that must remain dormant may be activated only to find that the surrounding environment has been altered to the extent that its code no longer makes any sense within the language with which it is supposed to interact. A well-defended system will be periodically patched so that changes like this should be routine. The result is that while militaries may spend years embedding capabilities into a network, they have no guarantees that when the time comes it will function as anticipated. Consequently, cyber attacks are unpredictable in their effects and are difficult to synchronise with established military planning and operations. This is a significant limitation, as the high lethality, mobility and relatively low quantities of most modern military capabilities mean that precise timing and coordination between different joint force elements are even more essential than they were in previous eras.

The Military Application of Cyber Weapons

There are limited scenarios in which a cyber attack can have a decisive operational or strategic effect when used in isolation. The most famous strategic cyber attack is the Stuxnet worm which formed part of a series of attacks on Iranian nuclear fuel enrichment centrifuges from 2008.[24]

---

23 Jens Stoltenberg, 'NATO Will Defend Itself', NATO, last updated 29 August 2019, <https://www.nato.int/cps/en/natohq/news_168435.htm?selectedLocale=en>, accessed 30 March 2021.

Estimates vary but some suggest that enough disruption was caused to set back Iranian nuclear enrichment by 18 to 24 months. However, the system reconnaissance and infiltration process took years, having reportedly begun in 2005.[25] Furthermore, when Stuxnet leaked accidentally beyond the air-gapped nuclear facility network, its code was subsequently identified and analysed by states and cyber security firms around the world. This resulted in public attribution of the Stuxnet attack itself to intelligence agencies in the US and Israel, as well as the probable attribution of other cyber weapons which appear to function in a similar manner.[26] It also gave Iran and other national and civilian organisations the opportunity to examine a strategic cyber payload described by a senior cyber security professional as 'by far the most complex piece of code that we've looked at – in a completely different league from anything we'd ever seen before'.[27]

The Stuxnet leak, and the US/Israeli cyber operations against Iran's uranium enrichment programme as a whole, illustrate some of the key practical characteristics of cyber warfare as a standalone tool of national power. Attacks on defended systems require years of preparation to identify, infiltrate and exploit weak points. Once triggered, even very sophisticated payloads will be discovered, analysed and ultimately identified. The effects created can only ever be temporary in isolation, as states will patch their systems and replace damaged machinery if not prevented from doing so by other means. When the payloads are ultimately identified and analysed, peer and near-peer states can use insights gained to ascribe probable attribution, and also potentially to develop or enhance their own offensive cyber capabilities.

Less sophisticated cyber payloads than Stuxnet-type cyber weapons can still be highly effective in causing standalone disruption and potentially physical damage if conducted against less defended systems such as those which control civilian infrastructure. Examples include the attack on a German steel mill in 2014, and the WannaCry ransomware attacks which caused major disruption to the NHS in 2017.[28] States such as Russia and

---

[24]   David E Sanger, 'Obama Order Sped Up Wave of Cyberattacks Against Iran', *New York Times*, 1 June 2012.

[25]   *Ibid.*

[26]   Josh Fruhlinger, 'What is Stuxnet, Who Created It and How Does It Work?', *CSO Online*, 22 August 2017, <https://www.csoonline.com/article/3218104/what-is-stuxnet-who-created-it-and-how-does-it-work.html>, accessed 3 March 2021.

[27]   *Ibid.*

[28]   TrendMicro, 'German Steel Plant Suffers Significant Damage from Targeted Attack', 12 January 2015, <https://www.trendmicro.com/vinfo/fr/security/news/cyber-attacks/german-steel-plant-suffers-significant-damage-from-targeted-attack>, accessed 3 March 2021; Josh Fruhlinger, 'What Is WannaCry Ransomware, How

North Korea regularly use quasi-criminal groups of hackers operating under loose state direction to demonstrate to expert communities their ability to disrupt and even directly damage civilian industry and critical national infrastructure (CNI). This capability is then tacitly leveraged to enhance deterrence messaging against potential state opponents. Some of these attacks exploit zero-day vulnerabilities, but most employ simple massed phishing attacks to infiltrate potential target networks. As such, they are much more effective against networks which rely on commercial operating systems which are well understood, do not use three-factor authentication security procedures and may be slow to implement security patches. The purpose of most attacks of this nature is also far less specific than strategic cyber weapons such as Stuxnet; either theft of information across a broad industrial sector, extortion or simply disruption for its own sake.

For most Western states, there are significant legal and ethical barriers to the use of offensive cyber capabilities against civilian targets, at least in any overt attempt at deterrence messaging. In any case, the ability to infiltrate civilian infrastructure and cause temporary disruption or even limited destructive effects is unlikely to deter a state opponent from conventional military action if it has already committed to such a high-risk course of action. The recent conflicts in eastern Ukraine and Nagorno-Karabakh provide compelling evidence that states are prepared to suffer extensive damage to both civilian and military targets without backing down from kinetic confrontations where national interests are clearly defined. As such, the immediate question for those who see the apocryphal cyber capability to 'turn out the lights in Moscow' as a cheaper replacement for conventional and nuclear deterrence tools is: what effect would such an attack have on a determined adversary, and for how long? It is perhaps ironic that hyperbolic narratives, such as that advanced by Nicole Perlroth,[29] use Ukraine as the example of how devastating cyber attacks can be, while somehow overlooking the extent to which Ukraine has weathered and recovered from Russia's cyber onslaught.

Whilst there are few situations in which cyber weapons can replace more traditional forms of military capability outright, there are many potential ways in which offensive cyber tools can be used as part of joint operations. For example, a previously inserted cyber payload might be

---

Does it Infect, and Who Was Responsible?', *CSO Online*, 30 August 2018, <https://www.csoonline.com/article/3227906/what-is-wannacry-ransomware-how-does-it-infect-and-who-was-responsible.html>, accessed 3 March 2021.

[29] Nicole Perlroth, *This Is How They Tell Me the World Ends: The Cyberweapons Arms Race* (London: Bloomsbury, 2021).

used to disable or disrupt key adversary defence systems at a critical moment to enable a conventional attacking force to physically bypass or destroy them. Israel allegedly carried out just such an attack against the Syrian air defence network in 2007.[30] The normally bristling air defence network was rendered blind to an Israeli Air Force strike package which entered Syrian airspace, bombed a nuclear facility and withdrew unchallenged.[31] It is still unclear in the public domain whether this was a case of a cyber payload or advanced electronic warfare techniques opening the door for the Israeli aircraft. In general terms, whereas cyber attacks involve the theft, alteration or deletion of data, electronic warfare involves the detection or disruption of enemy sensors and the transmission of data using electromagnetic energy. While both rely on an attacker having knowledge of how a target system operates, electronic warfare can be applied as an active effect in real time without lengthy network reconnaissance, infiltration and payload activation. In practice, the line between the two techniques is increasingly blurring. For example, information about the exact software and hardware configurations of a radar, obtained via cyber attacks against the defence network which it sits within, might be used to tailor sophisticated electronic warfare attack options using digital emitters such as active electronically scanned array radar sets. In this context, however, the cyber attack would have occurred long before the electronic attack, with the theft of data enabling conventional electronic warfare capabilities. What is clear, however, is that the Israeli attack in 2007 was successful because it blended cyber and/or electronic attack with a precisely choreographed application of traditional firepower. The attack on the Syrian air defence network itself was not intended to create strategic effects, but as an enabling capability it greatly reduced the risk profile and increased the effectiveness of an otherwise conventional strike operation.

The use of cyber attacks to 'open the door' for conventional forces in this way is dependent on several key factors. First, that the potential target can be identified well in advance of any requirement to exploit a breach. This is to allow sufficient time for the required reconnaissance, penetration and delivery stages of cyber attack preparation to be conducted. Second, in the case of an attack against a target such as an integrated air defence system (IADS), this activity will have to be coordinated with traditional intelligence gathering about the physical

---

[30] Sharon Weinberger, 'How Israel Spoofed Syria's Air Defense System', *WIRED*, 4 October 2007.
[31] *BBC News*, 'Israel Admits Striking Suspected Syrian Nuclear Reactor in 2007', 21 March 2018.

locations, individual system capabilities and patterns of movement exhibited by the radars and missile launchers themselves. This is so that the required virtual effects can be identified in terms of practical effects in the physical dimension, and it requires in-depth intelligence which will need to be gathered in multiple domains. This in turn introduces a third dependency for using cyber weapons in this way to enable conventional attacks; permissions and security clearance levels. Continuing with the IADS example, the required physical and network reconnaissance and intelligence-gathering activities would likely draw on resources from air force(s), multiple intelligence agencies and special forces operations. Many of these assets will require very high-level authorisation for tasking, and the products generated will be highly classified and not widely distributed within the military as a whole. As such, the successful use of cyber attacks as a coordinated enabler for conventional military operations requires operational planners to be aware of the art of the possible, in areas where they are unlikely to be in most countries by default. The final major dependency for this sort of attack is a linked one. Imagine a case where a cyber capability has been successfully developed and implanted with a compatible activation mechanism in a specific adversary defence system, and an operational commander is aware of it. In order to employ it, that commander must be able to justify its use to enable their conventional operation, given that once the cyber weapon is used the adversary will rapidly reset their systems, discover the vulnerability and patch it. In the case of the 2007 attack by the Israeli Air Force on a Syrian nuclear reactor, head of state authorisation would almost certainly have been granted for the operation, and the expending of a developed cyber capability (if indeed it was one) embedded over several years in the Syrian IADS deemed worthwhile to enable it.[32] However, for more routine or ongoing military operations, cyber weapons will likely remain a capability held for specific situations rather than a regular feature. Furthermore, if military operations take place in an unforeseen context, it is unlikely that sufficient time will be available to prepare as thoroughly as Israel was able to do against its traditional rival to the north.

A more ad-hoc use of disruptive cyber operations against CNI and military enabling capabilities is more likely to characterise the routine use of cyber capabilities in warfare in the coming decade. If the requirement is not to create a catastrophic failure in key military systems at a specific point in time to directly enable conventional operations, then the

---

[32] Ronen Bergman, *Rise and Kill First: The Secret History of Israel's Targeted Assassinations* (New York, NY: Random House, 2018), pp. 590–94.

dependencies placed on cyber specialists are less onerous. For example, Russia has conducted a range of cyber attacks against Ukrainian military and civilian systems as part of its hybrid war against the country since 2014. These have included attacks on Ukrainian military communications in coordination with intensive electronic warfare operations, as well as attacks on the national power grid and banking system. The result was to impose further disruption and costs on the government in Kyiv in conjunction with sustained direct and indirect military support to rebel groups in Donetsk and Luhansk oblasts. In other words, cyber warfare offers states another vector through which to pressure opposing states in conjunction with traditional hard and soft power tools. In a NATO context, obvious potential targets for Russian cyber attacks in the event of a major crisis or limited conflict would be the networks which control the civilian port and rail infrastructure which would be critical for mobilisation and the deployment of conventional military forces to eastern Europe. Such attacks could have significant or even decisive effects on the strategic picture for the Alliance if coupled with sustained conventional and nuclear force posturing. However, on their own, they would only delay rather than prevent military capabilities being brought to bear.

In considering the impact of cyber warfare on future operations, militaries are clearly correct to be taking the security of their systems seriously. Offensive cyber operations offer opportunities to gain contextual tactical advantage by disrupting the functioning of enemy systems and undermining the adversary's confidence in their equipment. They also allow for the shaping of the human terrain around which fighting takes place when aimed at civilian systems. Nevertheless, cyber attacks should not be seen as an alternative to hard power. Nor can cyber attacks function as a standalone solution.

Finally, the opportunities offered by offensive cyber capabilities are difficult to guarantee at a specified time and place, and so cannot be relied on as a foundational capability in deterrence terms. Cyber warfare is real and cannot be ignored. But for now, it remains slow and imprecise compared to more traditional levers of military power.

II. THE GREY ZONE IS DEFINED BY THE DEFENDER

SIDHARTH KAUSHAL

Over the course of recent years, it has become common to hear the term 'grey-zone strategy' invoked to conceptualise the ways in which a range of state competitors are pursuing revisionist goals. Per this understanding, Western adversaries such as Russia, China and Iran have become adept at using a range of tools short of open warfare to challenge the status quo, leaving Western policymakers scrambling to come up with appropriate and proportionate responses.[1] Similarly, the US and its allies have their own grey-zone tools, including support for colour revolutions, financial sanctions and cyber attacks. Thus, for example, the US Joint Chiefs of Staff envision an operating environment defined by sub-threshold activity and attendant ambiguity.[2] Nor is this an exclusively American view: the UK's Integrated Operating Concept notes that 'our adversaries use an array of capabilities including their militaries below the threshold of war and in ways that challenge our political and legal norms'.[3] Similarly, France's 2017 national security strategy notes the challenge posed by legally ambiguous forms of aggression in the information space: 'Ambiguous postures and covert aggression are also becoming more common, with certain states making an increasing use of a wide variety of proxies.'[4]

---

[1] See, for example, Michael J Mazarr, *Mastering the Gray Zone: Understanding a Changing Era of Conflict* (Carlisle, PA: US Army War Studies Press, 2015).
[2] Joint Chiefs of Staff, 'Joint Operating Environment 2035: The Joint Force in a Contested and Disordered World', 14 July 2016, <https://www.jcs.mil/Portals/36/Documents/Doctrine/concepts/joe_2035_july16.pdf?ver=2017-12-28-162059-917#:~:text=The%20Joint%20Operating%20Environment%202035,and%20its%20allies%20in%202035>, accessed 27 October 2021.
[3] Ministry of Defence (MoD), 'Integrated Operating Concept', August 2021, p. 8.
[4] Republic of France, 'Defence and National Security Strategic Review 2017', 2017, p. 47.

The use of political subversion, unbadged 'little green men', cyber attacks and the deployment of paramilitary units to prosecute territorial aims have all been described as instances of a supposedly new model of grey-zone warfare. Revisionist states have, it is contended, found ways to circumvent Western conventional strengths by exploiting the cumulative effects of persistent competition short of open warfare.[5] This view was captured by General Joseph Dunford, then chairman of the Joint Chiefs of Staff, when he claimed that adversary sub-threshold competition exploited an American mindset that drew a clear distinction between peace and war, and which lacked the conceptual architecture to frame competition.[6]

This chapter contends that the distinction between sub-threshold grey-zone tools and the tools used in conventional warfare is analytically unhelpful and can distort strategic decision-making. First, defining the concept of the sub-threshold space in negative terms – grouping together all forms of competition that do not entail high-intensity warfighting – adds little analytical value. Second, this distinction omits the relationship between a state's posture for high-intensity warfighting and its ability to compete at lower levels of intensity. Finally, defining the competitive space in terms of instruments used and the implicit assumption that certain tools are inherently ambiguous obscures the agency that both competitors have in delineating boundaries and the critical importance of what Herman Kahn dubs the 'systems competition' to define the thresholds at which competition enters a different phase.[7] Ultimately, the threshold between war and peace is not absolute, and is defined by the defender.

The Pitfalls of the Grey-Zone Concept

Central to the concept of the grey zone is the notion of ambiguity. It has been argued that means that cannot be unambiguously defined as an act of war can serve as tools by which the status quo can be altered without recourse to high-intensity warfighting. Following this argument, events such as the Russian annexation of Crimea, China's militarisation of the South China Sea, and Iran's use of mining in the Strait of Hormuz and missile strikes against Abqaiq and Khurais all challenge Western concepts

---

[5] Mazarr, *Mastering the Gray Zone*, pp. 10–20.
[6] Colin Clark, 'CJCS Dunford Calls for Strategic Shifts; "At Peace or at War Is Insufficient"', *Breaking Defense*, 21 September 2016.
[7] Herman Kahn, *On Escalation: Metaphors and Scenarios* (London: Routledge, 1965), pp. 236–50.

of deterrence by not presenting a clear *casus belli*.[8] Plausible deniability supposedly enabled the territorial or political status quo of each revisionist state's region to be challenged without an open declaration of war. Other analysts have also nested political subversion, economic coercion and cyber attacks within the rubric of grey-zone warfare.[9]

Three considerations are worth bearing in mind here. First is the risk of the concept serving as an analytical dustbin – a general purpose concept that encompasses a range of actions which share little in common other than the fact that they do not include high-intensity warfighting.[10] Kinetic attacks that constitute acts of limited warfare and activities which are more consistent with the conduct of peacetime statecraft such as economic coercion are lumped together under the rubric of grey-zone warfare.[11] Concepts which define the criteria for inclusion in negative terms as opposed to shared characteristics of their subjects do more to confound than to clarify. The risk is that dissimilar competitive strategies are grouped together simply because they fall short of warfare. Consider two recent examples of actions grouped under the rubric of grey-zone warfare: China's probing of Taiwanese defences and Iran's mining of tankers in the Persian Gulf. Each state's competitive strategy follows a distinct logic. Chinese probing appears to be aimed at the exhaustion of Taiwan's material capabilities. The costs of interception – $900 million a year – have caused the Taiwanese government to desist from sortieing aircraft to intercept People's Liberation Army aircraft.[12] This contributes to the incremental assertion of control over a territory's air and sea space that China wishes to annex as a stated policy. By contrast, Iran's approach had materially limited effects and was geared to exploit the

---

[8] Thomas Trask, Jonathan Ruhe and Ariel Cicurel, 'Countering Iran's Gray Zone Strategy', *RealClearDefense*, 18 October 2019, <https://www.realcleardefense.com/articles/2019/10/18/contesting_irans_gray_zone_strategy_114798.html>, accessed 1 March 2021; Andrew S Erickson and Ryan D Martinson (eds), *China's Maritime Gray Zone Operations* (Annapolis, MD: Naval Institute Press, 2019); David Carment and Dani Belo, 'Gray-Zone Conflict Management: Theory, Evidence, and Challenges', *Journal of European, Middle Eastern and African Affairs* (Vol. 2, No. 2, Summer 2020), pp. 21–41.

[9] Elisabeth Braw, 'Modern Deterrence: Preparing for the Age of Grey-Zone Warfare', *RUSI Newsbrief* (Vol. 38, No. 10, 5 November 2018).

[10] For more on the issue of conceptual clarity and the criteria by which it can be determined, see John Gerring, *Social Science Methodology: A Critical Framework* (Cambridge: Cambridge University Press, 2012), pp. 339–50.

[11] For example, the study of economic coercion has a long history as a field of coercion distinct from warfare. See David A Baldwin, *Economic Statecraft*, 4th edition (Princeton, NJ: Princeton University Press, 2020).

[12] *Reuters*, 'Taiwan Says Has Spent Almost $900 Million Scrambling Against Chinese Jets This Year', 7 October 2020.

psychologically magnifying effects of international market responses to limited stimuli in order to bring the West back to the negotiating table on sanctions. The fact that these strategies did not involve major war does not mean that they share very much in common. The risk of using the term 'grey-zone strategy' is that it overlooks a plethora of strategies that states may pursue in the competitive space, including the material erosion of an opponent's capabilities, costly signalling and territorial revisionism, among others.

A second concern is the assumption that certain instruments are inherently ambiguous or fall below a target state's escalatory thresholds. There is nothing inherent in a certain tool which makes it intrinsically non-escalatory. Consider, for example, the way in which Japan's attack on Pearl Harbor was prompted by the oil sanctions imposed by the Roosevelt administration. In other instances, by contrast, kinetic clashes did not lead to broader escalation, as was the case during the Korean War when Soviet pilots flew sorties against the US Air Force under North Korean colours.[13] In this case, as with many others, attribution was accomplished relatively quickly but a mutual desire to avoid broader escalation led both sides to downplay the fact that they were in combat. This is not to necessarily deny that some actions are easier to conduct on a covert basis than others, but rather that the key factor in defining precisely where the boundaries of the grey zone are is a conscious choice by the defender. This choice depends on several factors, such as the ability of the attacker to raise the risks of escalation which might occur in the event that a state of war is openly acknowledged and the trajectory of the overarching relationship between the two powers. States in a restrained competitive relationship comparable to the years of détente, for example, have incentives to broaden the boundaries of the grey zone.

A major risk inherent to defining strategies in terms of means used as opposed to ends sought is a misappreciation of the dynamics of risk and escalation. These are primarily a function of the significance of the political concession that one seeks from an opponent. For example, while many treated Iran's use of kinetic force in the Gulf in response to US sanctions as highly escalatory, there is no *a priori* reason to believe that recourse to kinetic assets represents a threshold by which escalation should be identified. The overarching aim of the maximum pressure strategy – a full-scale reassessment of Iranian foreign policy at a minimum and possibly even regime change – made escalation a foregone

---

[13] Austin Carson, *Secret Wars: Covert Conflict in International Politics* (Princeton, NJ: Princeton University Press, 2018).

conclusion, irrespective of whether economic or military instruments were used.

In a similar vein, when analysts speculate about whether a catastrophic cyber attack could be treated as the basis for kinetic action against a rival, they mistakenly conflate the dynamics of escalation with instruments used. If an opponent's strategy involves them seeking the significant disruption of one's own society – to a degree comparable to that which could be achieved by conventional warfighting – then this is likely to occur in the context of them seeking maximalist political objectives and would represent an escalation that few states would not treat as a justification for war, irrespective of the tools used to deliver this end. The instruments used to deliver a strategic objective are, while not insignificant in every case, of secondary importance to the nature of the objective itself.

Moreover, the assumption that certain tools are inherently ambiguous overlooks the way in which states can use the full spectrum of tools at their disposal – including warfighting assets – to set the boundaries of the competitive space in a way that gives them the advantage. Take, for example, the rationale for the expansion of the Soviet military, and particularly its navy under Sergei Gorshkov. The rationale provided by Soviet naval leaders, and accepted by the Kremlin, was that without a major navy any competition with the US at distance could escalate into a situation comparable to the Cuban missile crisis, where the Soviet Union – having engaged in sub-threshold competition via the emplacement of weapons in Cuba – was faced with either general warfare or an embarrassing climbdown by virtue of its naval inferiority and the nuclear balance.[14] In effect, if the Soviet Union wanted to compete in the sub-threshold space, where it believed it had advantages due to its network of proxies and allies as well as the allure of Communism in the developing world, it needed to have a credible conventional deterrent at reach as well as nuclear parity. Without this, the Soviet Union risked being threatened with escalation to localised conflicts which it could not win by a US that had an incentive to use its advantages above the threshold of warfare to limit what could be considered part of the sub-threshold space. Even if the US likely did not wish for conflict with the Soviet Union, its conventional and nuclear superiority meant that the Soviets had more to lose from escalation and

---

[14] David Holloway, 'Military Power and Political Purpose in Soviet Policy', *Daedalus* (Vol. 109, No. 4, 1980), pp. 13–30; Michael MccGwire, 'The Evolution of Soviet Naval Policy: 1960–1974', in Michael MccGwire, Ken Booth and John McDonnell (eds), *Soviet Naval Policy: Objectives and Constraints* (New York, NY: Praeger Publishers, Inc., 1975), p. 537.

greater incentives to take threats to escalate seriously even if they judged there to be a mutual disincentive to warfare. In other words, however unlikely a large-scale clash between the two powers, the conventional and nuclear balance of forces was an 'elephant in the room' which delineated what each party could risk doing in the competitive space.

Similarly, Russia's annexation of Crimea and China's creeping militarisation of the South China Sea were underpinned by escalation dominance over the immediate targets of revisionism and conventional deterrence, if not dominance, against likely external interveners.[15] Russian exercises conducted in tandem with the Crimean annexation had the effect of both fixing Ukrainian forces away from the theatre of action and raising the spectre of a wider conflict should Ukraine attempt to retake the peninsula. Similarly, the presence of the People's Liberation Army Navy (PLAN) over the horizon limits what local disputants in the South China Sea can do to constrain Chinese civilian, coastguard and maritime militia vessels. Moreover, external military intervention to reverse territorial gains made by each party is unlikely, less because of the inherent ambiguity of the actions taken than because of the military and political risks that this would entail.

The discourse surrounding the challenge of supposedly ambiguous tools ignores the many instances in which the use of supposedly grey-zone tools have generated either the threat or actual use of large-scale, conventional retaliation by the target state. For example, states such as Israel and Turkey have both threatened the use of force and taken action against opponents that lent support to sub-state proxies within their borders or housed elements that they deemed subversive.[16] Nor did the use of unbadged forces lead to any ambiguity in, for example, the 1999 Kargil War in which India, the target state, immediately identified these forces as regulars in plain clothes and treated Pakistani incursions as an act of war. Similarly, during the 1996 Taiwan Strait crisis, China's use of missiles tipped with dummy warheads under the aegis of conducting routine exercises in the waters near Taiwan produced a direct – if implicit – threat of a US response in the form of the dispatch of two carrier strike

---

[15] On the basic typology of strategy, see Robert J Art, 'To What Ends Military Power?', *International Security* (Vol. 4, No. 4, Spring 1980), pp. 3–35.

[16] Turkey threatened to invade Syria in 1998 unless the latter expelled Abdullah Ocalan and the Kurdistan Workers' Party leadership. See Nick Danforth, 'A Short History of Turkish Threats to Invade Syria', *Foreign Policy*, 31 July 2015. On Israel's responses to proxy subversion, see Wendy Perlman and Boaz Atzili, *Triadic Coercion: Israel's Targeting of States That Host Nonstate Actors* (New York, NY: Columbia University Press, 2019).

groups to the area.[17] Ambiguity, then, has rarely stayed the hand of states that believed they had viable conventional options.

Alternatively, when states have struggled to respond to low-intensity aggression, such as Ukraine in 2014 or the Philippines during the Scarborough Shoals crisis, their militaries were deterred from doing so effectively by the threat of escalation posed by large regular formations operating in coordination with proxies and paramilitaries.[18] In other words, the conventional balance of power often determines whether low-intensity aggression meets a large-scale kinetic response. When actors choose not to treat an act as one of open warfare, it is generally because they have been conventionally deterred or seek to prosecute a competition at a limited level of intensity as a policy choice.[19]

The Relationship Between Low-Intensity Coercion and Warfighting

The tendency, described above, to treat grey-zone activity as a means by which opponents exploit ambiguity to circumvent the conventional advantages of (primarily) Western states carries two risks:

1. Misperceiving adversary strategies and the relationships between low-intensity coercion and high-intensity warfighting.
2. Overlooking the agency that policymakers have in delineating the boundaries of the sub-threshold space to their own advantage.

To begin with the first risk, reconsider Colonel Harry G Summers Jr's analysis of the Vietnam War. The US, Summers presciently noted, had effectively misconstrued the nature of the war as being a conflict against proxy subversion by a North Vietnamese government which supported the Viet Cong. In truth, he noted, proxy subversion was an enabler. When the North finally absorbed the South in 1975, it was by a quintessentially conventional offensive. The Viet Cong had created the preconditions for it by forcing a US withdrawal, and it was this offensive that decided the political outcome of the war.[20] Had the US correctly

---

[17] J Michael Cole, 'The Third Taiwan Strait Crisis: The Forgotten Showdown Between China and America', *National Interest*, 10 March 2017.

[18] For a discussion of Russian actions, see Michael Kofman et al., *Lessons from Russia's Operations in Crimea and Eastern Ukraine* (Santa Monica, CA: RAND, 2017). On Chinese activities, see Sidharth Kaushal and Magdalena Markiewicz, 'Crossing the River by Feeling the Stones: The Trajectory of China's Maritime Transformation', *RUSI Occasional Papers* (October 2019).

[19] Carson, *Secret Wars*.

[20] Harry G Summers Jr, *On Strategy: A Critical Analysis of the Vietnam War* (New York, NY: Random House, 1995).

perceived that conventional forces were the North's centre of gravity, it might have pursued an approach comparable to that of the Easter Offensive in which conventional air power was used to support the Army of the Republic of Vietnam in neutralising North Vietnamese Army main force units and deter subsequent conventional attacks.[21] This would likely have been consistent with the public's willingness to bear costs and would have effectively neutralised the conventional force without which no amount of proxy subversion could deliver the ultimate aim of unification. The emphasis on the pacification effort in the South, and the immense costs it entailed, were the product of treating subversion as a novel strategy to be countered as opposed to an enabler for an eventual conventional clash.[22]

To use a more contemporary example, China's militarisation of small islands in the South China Sea has salience in the context of its doctrine of being able to fight 'local wars under informatized conditions'.[23] The ISR and sea denial assets on these islands can provide the PLAN with local overmatch in the South China Sea against most plausible opponents in circumstances short of protracted warfare. It is in their capacity to reinforce a strategy based primarily around conventional warfighting – and the impact that this has on the region's balance of power – that these developments have greatest salience. Similarly, in both Crimea and the Donbas, Russian forces achieved conventional overmatch against their Ukrainian counterparts – often in very bloody confrontations, as in the case of the Donbas conflict. Even if conventional forces are not used but are simply postured to limit an opponent's options – as was the case in Crimea – the conventional balance of forces is often central to any strategy. This approach uses a combination of tools which allows an actor to determine the rules of competition by creating an asymmetry of risk should a conflict escalate.[24]

---

[21] On the Easter Offensive, see Phil Haun and Colin Jackson, 'Breaker of Armies: Air Power in the Easter Offensive and the Myth of Linebacker I and II in the Vietnam War', *International Security* (Vol. 40, No. 3, 2016), pp. 139–78.

[22] On the historical parallels between the emphasis of the Kennedy and Johnson eras on new tools of strategy and the current grey-zone discourse, see John Lewis Gaddis, *Strategies of Containment: A Critical Appraisal of American National Security Strategy During the Cold War* (Oxford: Oxford University Press, 2005), chapters 7 and 8.

[23] See for example, M Taylor Fravel, 'China's New Military Strategy: "Winning Informationized Local Wars"', *China Brief* (Vol. 15, No. 13, 2 July 2015).

[24] For further information, see Sidharth Kaushal and Peter Roberts, 'Competitive Advantage and Rules in Persistent Competitions', *RUSI Occasional Papers* (April 2020).

This leads to the second challenge of framing the grey zone in terms of the inherent ambiguity of the tools used by a competitor – ignoring the extent to which defining the boundaries of the competitive space is a conscious policy choice that depends on the outcome of a parallel competition to set the rules of competition and to define the contours of precisely where the competitive space transitions to warfighting. Actors may have incentives to attempt to define the competitive space in narrow or broad terms, depending on their circumstances. A conventionally weaker actor has obvious incentives to broaden the scope of the competitive space and limit the possibility of high-intensity warfighting. Take, for example, the dynamic between Pakistan and India, where the two sides are effectively engaged in a competition to define whether proxy warfare is a sub-threshold activity. Pakistan's use of battlefield nuclear weapons is effectively geared to ensuring that this remains the case, while the Indian Cold Start doctrine entailing limited conventional offensives is meant to provide India with usable conventional options to respond to a proxy attack. Similarly, the assassination of Qassem Soleimani in response to proxy attacks on US personnel in Iraq was notable in the way it reframed the nature of the competition by demonstrating that a direct and potentially highly escalatory kinetic military response against a sponsoring state might occur as a result of proxy attacks – thus using conventional forces to narrow the boundaries of the sub-threshold competitive space for Iran. Defining some tools as inherently belonging to the grey zone leads policymakers to overlook the degree to which defining the boundaries of the competitive space is a strategic choice that must itself be based on an assessment of national strengths and the use of the levers of national power.

Conclusions

The treatment of low-intensity conflict, as well as the use of other tools of statecraft within a state's grand strategy, has some pedigree and is a useful subject of enquiry. That said, the emphasis on ambiguity evinced by both analysts and policymakers' discussions of the grey-zone concept does more to obscure than clarify.

First, the distinction between grey-zone actions short of war and warfighting is analytically unhelpful and obscures the role of kinetic action in many of the instances of revisionism grouped under the grey-zone rubric. While it is useful and necessary to talk about the strategies that states can pursue in the context of long-term competition, describing competition in the grey zone as a 'strategy' adds little analytical value.

Second, and finally, an overemphasis on ambiguity ignores the importance of Kahn's 'systemic competition' to define where the boundaries of the competitive space are, as well as the agency that the target state has in defining the contours of the grey zone. This has the effect of eliding the role of a state's conventional force posture and even its nuclear assets in the sub-threshold space, as it is the posturing of these assets that often delineates the boundaries.

III. DOING LESS WITH LESS IN THE LAND DOMAIN

NICK REYNOLDS

It is a long-established point of pride for smaller, lighter professional military forces that they can match larger, heavier forces. The US Marine Corps (USMC) has often considered itself as 'doing more with less'[1] when compared to the US Army, equivalent to the UK defence cliché of 'punching above our weight'.[2] Since the end of the Cold War, successive events have pushed most Western militaries to become smaller. The idea of a peace dividend was followed by attempts to make defence more efficient, and despite a brief trend of modest expansion during the War on Terror, this has been followed by further contraction. The official rationale usually involves efficiency and cost-effectiveness. Recently, an additional element of the debate has been brought to the fore: whether older, heavier platforms are survivable in the face of the increasing range, precision and lethality of offensive weapons and technology.

The UK's Integrated Review in 2021 required government policy to directly address these questions, and the result for the most part favoured smaller and lighter land forces. It framed the shrinkage of the British Army as a positive step, stating that 'the Army of the future will be leaner, more lethal, nimbler, and more effectively matched to current and future threats' while proposing personnel cuts 'from the current Full Time Trade Trained strength of 76,000 to 72,500 by 2025'.[3] The necessity of rectifying prior funding discrepancies by difficult prioritisation decisions was

---

[1] Raymond Priest, 'Doing More with Less', *Marine Corps Gazette* (Vol. 74, No. 10, 1990).
[2] Thomas Colley, 'Britain's Public War Stories: Punching Above Its Weight or Vanishing Force?', *Defence Strategic Communications* (Vol. 2, 2017), pp. 172–73.
[3] Ministry of Defence (MoD), *Defence in a Competitive Age*, CP 411 (London: The Stationery Office, 2021), p. 53.

certainly a major factor.[4] However, Chief of the General Staff General Sir Mark Carleton-Smith has argued that the Integrated Review left the British Army 'right-sized'.[5] Meanwhile, the Chief of Defence Staff focused his own assessment on the UK's technological capabilities and the benefits of structural changes geared towards competition over warfighting when justifying the cuts to the number of main battle tanks (MBTs) to be modernised and the total loss of the infantry fighting vehicle fleet.[6] Recently, the UK Defence Secretary claimed when proposing personnel cuts that 'the Army's increased deployability and technological advantage will mean that greater effect can be delivered by fewer people'.[7] The restructuring reflects serious thinking about the implications of technological change and how to maximise the utility of the armed forces.

While the need for modernisation is real, the focus on efficiency gains justifying cuts to the size of conventional forces risks conflating some quite different issues. That some tasks can be carried out with fewer personnel does not mean that all tasks demanded of land forces do not require mass. Reductions in mass are not equally consequential across the force, and those below certain levels begin to significantly constrain the ground a force can contest and the risks available to a commander irrespective of how lethal or enabled individual force elements are. It is important that the mantra of doing more with less is challenged, because increased capabilities enabled by specific technologies risk concealing very real reductions in the breadth of tasks a force can deliver once it is cut below certain critical thresholds. A focus on the expense of, and problems with, unmodernised, unrestructured military forces has proved to be deeply unhelpful, as has the narrative that new technologies should replace rather than complement legacy platforms. This chapter seeks to contend that far from maintaining capability, recent cuts to the British Army have

---

[4] Guy Anderson et al., 'The UK's Integrated Review and Defence Command Paper', *Janes*, 23 March 2021; Mark Lyall Grant, 'The Integrated Review's Concept of Global Britain – Is It Realistic?', King's College London, 19 July 2021, <https://www.kcl.ac.uk/the-integrated-reviews-concept-of-global-britain-is-it-realistic>, accessed 8 August 2021.

[5] RUSI, 'Land Warfare 2021: Welcoming Remarks and Opening Keynote' (14:24), 17 June 2021, <https://www.youtube.com/watch?v=JhcGUoNo6Hk>, accessed 6 August 2021.

[6] MoD and Nick Carter, 'Chief of Defence Staff Speech RUSI Annual Lecture', 17 December 2020, <https://www.gov.uk/government/speeches/chief-of-defence-staff-at-rusi-annual-lecture>, accessed 8 August 2021.

[7] MoD and Ben Wallace, 'Defence Secretary Oral Statement on the Defence Command Paper', 22 March 2021, <https://www.gov.uk/government/speeches/defence-secretary-oral-statement-on-the-defence-command-paper>, accessed 8 August 2021.

left it unable to effectively conduct operations that the Integrated Review suggests are critical to the UK's foreign policy strategy.

Not All Reductions in Mass Are Equal

Before exploring the problems with force reductions, it is important to acknowledge legitimate reasons for why militaries are letting go of some capabilities and fielding smaller formations. In the US, there is an ongoing public debate regarding the restructuring of the USMC, which is divesting itself of all its heavy armour and virtually all its tube artillery, from 21 battalions to five. As it also loses three marine infantry battalions and two amphibious vehicle companies, as well as two transport and two attack aviation squadrons and associated ground support, this constitutes more than a restructuring but a significant contraction of the force by 7%.[8] The future USMC will favour lighter forces to focus on the littoral environment of the South and East China Seas,[9] representing perhaps the most radical restructuring of any Western military at present. The planned changes are controversial, with concerns about this approach focusing on one problem set to the exclusion of other possibilities.[10] However, the proposed force structure is the result of dedicated analysis to specific operational tasks against which the force intends to optimise.

---

[8] Andrew Feickert, 'New U.S. Marine Corps Force Design Initiatives', Congressional Research Service, IN11281, updated 2 March 2021, pp. 1–2; Gina Harkins, 'Marines to Shut Down All Tank Units, Cut Infantry Battalions in Major Overhaul', *Military.com*, 23 March 2020, <https://www.military.com/daily-news/2020/03/23/marines-shut-down-all-tank-units-cut-infantry-battalions-major-overhaul.html>, accessed 29 October 2021.

[9] Megan Eckstein, 'New Marine Corps Cuts Will Slash All Tanks, Many Heavy Weapons As Focus Shifts to Lighter, Littoral Forces', *USNI News*, 23 March 2020, <https://news.usni.org/2020/03/23/new-marine-corps-cuts-will-slash-all-tanks-many-heavy-weapons-as-focus-shifts-to-lighter-littoral-forces>, accessed 29 October 2021; David B Larter, 'The US Marine Corps Wants Grunts Packing Deadly Swarming Drones', *Defense News*, 9 December 2020, <https://www.defensenews.com/naval/2020/12/09/the-us-marine-corps-wants-grunts-packing-deadly-swarming-drones/>, accessed 29 October 2021; Peter Ong, '"The U.S. Marine Corps Has Divested in Their Tanks" Well, What Does That Mean?', *Naval News*, 25 September 2020, <https://www.navalnews.com/naval-news/2020/09/the-u-s-marine-corps-has-divested-in-their-tanks-well-what-does-that-mean/>, accessed 29 October 2021.

[10] Mallory Shelbourne, 'Panel: New Focus on China Fight Could Rob Marine Corps of Versatility', *USNI News*, 30 July 2020, <https://news.usni.org/2020/07/30/panel-new-focus-on-china-fight-could-rob-marine-corps-of-versatility>, accessed 29 October 2021; Mark Cancian, 'Don't Go Too Crazy, Marine Corps', *War on the Rocks*, 8 January 2020.

The French army has long favoured lighter, more mobile forces, having tailored their force structure to fit within the constraints of their limited logistics capability, and generally tolerate greater risks.[11] As the general officer commanding its Centre for Doctrine and Command Teaching, Michel Delion noted, 'the French army has maybe a less technophile approach than other armies'.[12] However, its current modernisation trajectory blends technological updates and multi-domain concepts that are broadly comparable to their US and UK equivalents, including a vehicle upgrade programme for its heavy forces, with the divergent commitment to increasing spending by a projected 46% between 2018 and 2025[13] and to generating combat mass on land.[14] The Chief of the French Army Thierry Buckhard has notably committed to regenerating the army's ability to conduct combined-arms manoeuvres at division scale.[15] Nevertheless, French military concepts do expect smaller force packages to offer greater combat power through the application of reconnaissance, precision fires and manoeuvre.[16]

Aiming to do more with less is an appropriate approach under the right conditions. Persistent ISTAR and precision fires are an affordable means of achieving military effect and are playing a critical role in counterterrorist and counterinsurgency campaigns, and were recently seen as enabling a decisive military outcome in Nagorno-Karabakh. Meanwhile, lighter forces are easier to deploy and subsequently use in places where heavy forces cannot be deployed within a relevant

---

[11] Michael Shurkin, 'What It Means to Be Expeditionary: A Look at the French Army in Africa', *Joint Forces Quarterly* (Vol. 82, No. 3, 2016), pp. 76–78.

[12] Sydney J Freedberg Jr, 'Budget Up, French Army Preps for Major Wargames With US', *Breaking Defense*, 25 November 2020, <https://breakingdefense.com/2020/11/budget-up-french-army-preps-for-major-wargames-with-us/>, accessed 29 October 2021.

[13] *The Economist*, 'The French Armed Forces are Planning for High-Intensity War', 31 March 2021.

[14] Armée de Terre, 'Strategic Vision of the Chief of the French Army: "2030 Operational Superiority"', April 2020, pp. 2, 6, <https://franceintheus.org/IMG/pdf/french_army_strategic_vision_2020.pdf>, accessed 27 October 2021; Ben McLennan, 'Confronting a Foreboding Future: The French Army's Strategic Vision', *The Strategist*, 14 August 2020; Audrey Quintin, 'Progress on the Scorpion Program: France's Plan to Upgrade its Motorised Capacity', Finabel, 26 February 2020, <https://finabel.org/progress-on-the-scorpion-program-frances-plan-to-upgrade-its-motorised-capacity/>, accessed 29 October 2021; *Army Recognition*, 'French Army Accelerates Modernization of Land Force', 19 September 2019, <https://www.armyrecognition.com/september_2019_global_defense_security_army_news_industry/french_army_accelerates_modernization_of_land_force.html>, accessed 29 October 2021.

[15] Armée de Terre, 'Strategic Vision of the Chief of the French Army'.

[16] Guy Hubin, *Perspectives tactiques* (Paris: Economica, 2009).

timeframe. Moreover, counterarguments are often polemic, with NATO's primary adversary Russia, and China, the US's 'pacing threat',[17] held to a double standard. The Russian military's aim of shrinking from 1.35 to 1 million personnel[18] and its increased focus on precision fires, electronic warfare and drones is generally portrayed as a well-formulated and threatening restructuring effort while domestic restructuring along similar lines is disparaged. Restructuring efforts are a necessity, and those formations tasked with executing the tactical fight – the archetypal combat arms and manoeuvre elements that form the core of any professional military – can produce an outsized effect when backed up by modern enablers.

Much of the appeal of current debates and trends in force modernisation and restructuring are driven by the efficacy of the combination of pervasive, persistent ISTAR and long-range precision fires. Modern ISTAR in the form of drones can cue on precision fires, destroying even heavily protected targets. This can be done in a way that is affordable due to the comparatively low cost of drones and munitions, while still providing the ability to deliver lethal effect in depth and at scale.[19] As long-range precision fires do not need to saturate targets to destroy them, they do not need to be launched en masse – or at least not to the degree required in previous conflicts, albeit still requiring to be sufficiently distributed in order to benefit from dispersion and avoid being targeted and attrited by counter-battery fire. Furthermore, the offence–defence paradigm has shifted. Countermeasures have proven increasingly expensive to develop while still struggling to reliably protect platforms and formations. Fears about the lethality of future warfare triggered by the destruction of the Armenian army in Nagorno-Karabakh are not unfounded. Thus, there is a practical and psychological disincentive to retaining a large unmodernised force that is vulnerable to long-range fires; if a force will simply be destroyed while proving unable to adequately defend itself against a peer adversary, it has little deterrent value in high-end conflict, its use will be limited and it will not represent

---

[17] Jim Garamone, 'Milley Makes Case for U.S. Military Keeping Up With Global, Technology Changes', US Department of Defense, 2 December 2020, <https://www.defense.gov/Explore/News/Article/Article/2432855/milley-makes-case-for-us-military-keeping-up-with-global-technology-changes/>, accessed 29 October 2021.

[18] Defence and Military Analysis Team, 'Russia's Armed Forces: More Capable by Far, But For How Long?', *IISS Military Balance Blog*, 9 October 2020, <https://www.iiss.org/blogs/military-balance/2020/10/russia-armed-forces>, accessed 29 October 2021.

[19] Liam Collins and Harrison 'Brandon' Morgan, 'Affordable, Abundant, and Autonomous: The Future of Ground Warfare', *War on the Rocks*, 21 April 2020.

value for money unless it operates in an alliance context, providing support or combat mass to a more capable ally.

Regrettably, while policy documents have alluded to these changes in the character of warfare, cuts to UK land forces have been driven principally by fiscal calculations. The UK has experienced more extreme shrinkage than comparable militaries, even before the Integrated Review. The 2010 Strategic Defence and Security Review (SDSR) was replete with references to efficiency gains, cost-effectiveness and enhanced capability,[20] while the Ministry of Defence's (MoD) 2012 Defence Equipment Plan laid out similar aspirations of efficiency concurrent with investment and improved effectiveness.[21] Yet, while troop strength was reduced, no new major systems have entered service. The 2015 SDSR and 2016 Defence Equipment Plan raised awkward questions when it became apparent that the efficiency savings required to continue the planned level of investment in the armed forces outstripped the ability of the MoD and services to deliver them.[22] Furthermore, the 2017 National Security Capability Review shifted its language towards sustainment and resilience in the face of new threats.[23] Here, structural changes and budget cuts are driven more by financial constraints than any calculations about threats and capability. Problems have culminated to the point that in January 2021, the National Audit Office declared '4 years in a row that we have reported that the Equipment Plan has been unaffordable'.[24]

Modernisation is Not Minimisation

It is clear that conflicting imperatives allow rhetoric to be deployed to support contradictory propositions: that cuts and changes can either represent adaptation and efficiency, or the irresponsible cutting of core capability. Untangling debates around which interpretation is more accurate can be difficult, for there is a significant theory–praxis gap in modern warfare. Recent small-scale examples such as Nagorno-

---

[20] HM Government, *Securing Britain in an Age of Uncertainty: The Strategic Defence and Security Review*, Cm 7948 (London: The Stationery Office, 2010), pp. 10, 26, 33, 59, 69.
[21] MoD, 'The Defence Equipment Plan 2012', p. 4.
[22] Commons Defence Select Committee, 'Gambling on "Efficiency": Defence Acquisition and Procurement Contents: 3. The Defence Equipment Plan', 14 December 2017, <https://publications.parliament.uk/pa/cm201719/cmselect/cmdfence/431/43106.htm>, accessed 10 March 2021.
[23] HM Government, 'National Security Capability Review', March 2018, pp. 2–6, 14–17.
[24] National Audit Office, *The Equipment Plan 2020 to 2030*, HC 1037 (London: National Audit Office, 2021), p. 4.

Karabakh, which involved less-capable belligerents such as Armenia who lacked a modern IADS, provide very limited evidence sets.[25] NATO and Western-style militaries have not faced a peer adversary in fighting at scale for many years, and so the system of capabilities that modern militaries field have not been put to the test in adversarial conditions. While this lack of fighting is merciful, the result is a greater degree of uncertainty about how peer adversaries and Western forces will interact in practice. It is especially difficult for outsiders to determine the exact impact of structural changes to military forces, and public opinion is a poor gauge of whether given changes are warranted or sensible. Even well-intentioned restructuring efforts could prove difficult to rectify or reverse if core capabilities had been sacrificed for expected improvements in efficiency and capability which proved illusory. When misapplied, the rhetoric of change can therefore be disastrous. The idea of doing more with less, which is politically palatable and financially attractive in a resource-constrained environment, is dangerous for exactly this reason unless the government, military hierarchy and general public have a clear understanding of what they want to get from their investment in their military, what particular lines of investment buy them in terms of capability and where the inefficiencies within the military are to be found. The Russian military's contraction in terms of personnel numbers[26] occurs in the context of a force that still operates at a superior scale to its European adversaries and is growing in terms of defence spending.[27] Likewise, the USMC's shift away from heavy forces is explicitly designed to prevent duplication of effort with the US Army, rather than to divest the services as a whole of capability, and the USMC's new structure is self-consciously specific to the Pacific theatre.[28] Forces might have scope to become leaner, but still need to be able to operate at a relevant scale. This requires a degree of combat mass, and traditional calculi about the number of personnel, formations, combat vehicles and systems required in the context of a state's strategic circumstances remain relevant.

---

[25] Shaan Shaikh and Wes Rumbaugh, 'The Air and Missile War in Nagorno-Karabakh: Lessons for the Future of Strike and Defense', Center for Strategic and International Studies, 8 December 2020, <https://www.csis.org/analysis/air-and-missile-war-nagorno-karabakh-lessons-future-strike-and-defense>, accessed 28 August 2021.

[26] Defence and Military Analysis Team, 'Russia's Armed Forces'.

[27] Andrew Radin et al., *The Future of the Russian Military: Russia's Ground Combat Capabilities and Implications for U.S.-Russia Competition* (Santa Monica, CA: RAND, 2019), pp. 30–36.

[28] Ong, '"The U.S. Marine Corps Has Divested in Their Tanks" Well, What Does That Mean?'.

Take, for instance, the recent non-combatant evacuation operation from Afghanistan as a demonstration of the minimum necessary mass to achieve politically essential military tasks. The international coalition needed to secure Kabul airport and use it as a staging area for operations to recover their nationals and allies. This turned out to require the best part of an infantry division, including elements from the 82nd Airborne, US Marine Corps, the UK's 16 Air Assault Brigade, Turkish, French, German and smaller elements of other coalition forces.[29] Until available at this strength, the force was insufficient to prevent the runway being overrun by non-combatants.[30] This force was subsequently largely committed, holding back large crowds, providing medical assistance and screening evacuees. None of these tasks could be made more efficient by machines. Yet, this force was only sufficient because the Taliban had chosen not to attack coalition forces.[31] Had they put the runway under indirect fire, the cordon would have needed to be expanded, significantly increasing the size of the force necessary. Given that the UK's highly deployable force amounts to a single under-strength infantry brigade, the UK was dependent on the US presence until their withdrawal.[32] Had the UK bent itself out of shape to deploy more infantry and extend the presence beyond the US, they would have been reliant on the Taliban not escalating. Otherwise, the operation would suddenly have overstretched their capacity to maintain the perimeter. The UK's policy options were clearly constrained by what the US would enable, and the Taliban allow. The result was that, in spite of the commendable efforts of UK troops on the ground, nationals and many Afghans who the UK intended to evacuate were left in Kabul, under threat from the Taliban. The UK lacked the mass to secure a single piece of infrastructure abroad, and no amount of efficiencies within the force would have altered that calculus.

Regarding the UK's commitments to NATO, a division is 'the principal element of the British contribution to a coalition'.[33] Here, the UK's Warfighting Division is a useful case study for determining whether the UK operates at an appropriate scale to fulfil its stated commitments. It is worth emphasising that commitments to NATO remain at the heart of the

---

[29] Brief to RUSI by 16 Air Assault Brigade Combat Team on 12 October 2021.

[30] Luke Harding and Ben Doherty, 'Kabul Airport: Footage Appears to Show Afghans Falling From Plane After Takeoff', *The Guardian*, 16 August 2021.

[31] Rupan Jaim, 'Taliban Guards Continue to Provide Security Outside Kabul Airport Taliban Official', *Reuters*, 26 August 2021.

[32] Daniel Kramer, 'Afghanistan: Why Can't the UK Hold Kabul Airport Without the US?', *BBC News*, 27 August 2021.

[33] Centre for Historical Analysis and Conflict Research, 'The Big Picture: The UK Warfighting Division in Context', *Ares & Athena* (No. 6, November 2016), p. 6.

Integrated Review. The numbers are troubling. Only 148 of the British Army's 227 Challenger 2 MBTs are to be retained and upgraded after the Integrated Review,[34] and 3rd (UK) Division will operate 112 of these in two regiments.[35] Within the division, 1 Armoured Infantry Brigade, 12th Armoured Infantry Brigade and 20th Armoured Brigade[36] form a Deep Recce Strike Combat Team and two Heavy Brigade Combat Teams (BCTs). Each Heavy BCT will derive its principal combat power from an armoured regiment of MBTs, specifically the Type 56 Armoured Regiment, so-called because it consisted of 56 MBTs split into three 18-strong squadrons, a model that remains fundamentally unchanged from when the Army 2020 plan was developed.[37] While the UK's fleet management dictates that units do not routinely hold a full complement of MBTs, in the event of conflict around 112 out of 148 MBTs would sit within the Warfighting Division. By contrast, a standard-configuration Russian tank regiment, their closest equivalent formation to 3rd (UK) Division's Type 56 Armoured Regiment, would derive its principal combat power from 93 T72B3 MBTs that would be spread across three battalions of 31 MBTs each, supported by a motor-rifle battalion equipped with 41 BMPs (*Boyevaya Mashina Pekhoty*, meaning 'infantry fighting vehicles'). The two supporting motor-rifle regiments would each include another tank battalion, bringing the total number of MBTs in a Russian division to 155.[38] Qualitatively, Russian modernisation of their armoured forces make the workhorse T72B3 MBT a threat to Challenger.[39] UK doctrine dictates that a 3:1 ratio in the close battle should be sought to achieve decisive advantage, and is a common planning assumption. The numerical disadvantage between the platforms delivering combat power that the UK would face in a high-end warfighting scenario, and the lack of a second echelon to replenish losses when factoring in the need to retain a training fleet, indicate that

---

[34] MoD, *Defence in a Competitive Age*, CP 411 (London: The Stationery Office, 2021), p. 54.
[35] *Ibid.*, pp. 20, 68.
[36] British Army, 'Continual Operation Readiness: 3rd (United Kingdom) Division', <https://www.army.mod.uk/who-we-are/formations-divisions-brigades/3rd-united-kingdom-division/>, accessed 24 August 2021.
[37] British Army, *Transforming the British Army: Modernising to Face an Unpredictable Future* (London: The Stationery Office, 2012), p. 5.
[38] Konrad Muzyka, 'Russian Forces in the Western Military District', CNA, June 2021, p. 23, <https://www.cna.org/CNA_files/PDF/Russian-Forces-in-the-Western-Military-District.pdf>, accessed 24 August 2021.
[39] Will Flannigan, 'Facts Over Fear; T-14 Armata', *Wavell Room*, 19 February 2019, <https://wavellroom.com/2019/02/19/facts-over-fear-t14-take-the-threat-seriously/>, accessed 8 August 2021.

the UK's contribution to NATO is verifiably under strength. These deficiencies are often wished away with the assurance that the enemy would be attrited by higher-echelon effects before UK units are committed, but given that equivalent Russian formations also have significantly more artillery, and that the RAF would be unable to support ground units while conducting the suppression and destruction of enemy air defences, these assumptions are ill founded. Arguments that the character of war has changed and that novel capabilities invalidate these force ratios, while theoretically defensible, are themselves invalidated by the fact that no such novel capabilities are held within 3rd (UK) Division, nor is their acquisition at scale part of the UK's Equipment Plan. Thus, in light of the concept of operations of UK forces, the force is insufficient in scale to deliver against its stated task.

A related issue is the importance of enablement. As noted above, persistent ISTAR and precision fires are disproportionately influential in the current discourse, for this means of generating and delivering firepower is one area in which a smaller force can deliver large effects. Yet, in the land domain, these functions must support the ground manoeuvre units and other frontline combat elements. In the recent Nagorno-Karabakh conflict, the most important result of effective Azeri drone use was to attrit Armenian reinforcements and enablers such as artillery, allowing the Azeris to fight under advantageous conditions at the point of contact. The drones set decisive conditions but were still a shaping force. Closing with and destroying the enemy, physically manoeuvring and taking ground all had to be conducted much as before.

The same logic applies to non-kinetic measures. Much has been made of influence, cyber and psychological operations. In low-intensity conflict or other operations below the threshold of warfighting, these have been suggested as having the capacity to replicate the effect of kinetic measures without the expense, human cost or risk. Yet, these non-kinetic measures are most useful when they either enable ground manoeuvre forces to conduct their mission or amplify the effects produced by physical activity, an interactionist model between kinetic and non-kinetic effects that defies the narrative of replacement.

The Need for Skin in the Game

Interactionist approaches of both types, between both modernised enablers and traditionally structured and equipped ground combat forces, and kinetic and non-kinetic effects, are often used in the context of alliances and partnerships. Pioneered as a core element of campaigning by the US in 2001 in Afghanistan,[40] and continuing in comparable forms to this day, this has involved providing ISTAR and air assets to bolster local partner forces, with a small number of well-trained personnel providing a liaison function and working to integrate the disparate elements. Local partners provide combat mass enabled by modern firepower and targeting, and non-kinetic influence measures can be provided by whichever party is able to best tap into and leverage the relevant networks, with local partners dealing with local social groups and sub-state entities while patrons might manage influence in cyberspace and diplomatic initiatives with international stakeholders. While this approach originated for reasons of political sensitivities around presence and casualties, it is of use if the military providing high-end assets is too small to deploy at sufficient scale to achieve their aims, even if the political will to do so exists. In theory, high-end assets can provide sufficient capability to buy credibility as a partner or ally, and so partnered operations have been argued as a replacement for mass.

However, asking other partners and allies to provide the bulk of the ground forces, who accept the most physical risk and thus the vast majority of casualties when things go wrong, is problematic. A perceived unwillingness to commit blood as well as treasure in the eyes of partners and allies could be damaging to the state in question's reputation. Since the end of early, large-scale counterinsurgency deployments during the War on Terror, Western forces working with less-capable partners – as the US has done in Iraq and Afghanistan – have often been prevented from accompanying forces forward by political sensitivities about overinvolvement and casualties. Advisors have found ways of retaining credibility in the eyes of their partners under such circumstances, usually through demonstrating sufficient past combat experience coupled with bringing relevant and useful skills. Nevertheless, this has represented an ongoing challenge when one partner in a partnership is taking physical risks that the other is not.[41]

There is also the issue of the expertise that liaison personnel bring, for if the plugging in of ISTAR, air assets and other high-end capabilities includes the provision of the underpinning command-and-control (C2) structure, then advisors must be sufficiently experienced at working within a comparable system. Effective C2 partnering may prove difficult if the provider of C2 structures does not routinely operate at scale and lacks the relevant expertise.[42] Furthermore, as evidenced by the escape

---

[40] Yaniv Barzilai, *102 Days of War: How Osama bin Laden, al Qaeda & the Taliban Survived 2001* (Washington, DC: Potomac Books, 2013), pp. 100–18.

[41] Author interview with US officers advising the Afghan National Army, February 2020.

[42] *Ibid*.

of Osama bin Laden in 2001,[43] the ability of a handful of personnel, even in a coordinating role, to influence a partner force to fight in a desired manner towards specific operational objectives can be limited. The force providing the main ground presence will be able to determine the priorities and direction of operations at critical points. Relying on partners to do so inherently surrenders a great deal of strategic decision-making authority and an acceptance of enabling local partners to achieve their own objectives first and patron objectives second.

The evidence suggests that Western militaries should seek to retain their own ground combat mass and be willing to accept the risks associated with conducting operations in a sovereign manner. Furthermore, they should keep abreast of current technological trends and concepts of operation to ensure that they have an integrated range of capabilities. As survivability cannot be guaranteed at the level of small unit and individual platform defence due to the sheer lethality of modern offensive capabilities, this integration is critical. It is only by acting in concert through integration that different elements of a military force can remain situationally aware, coordinated, lethal and survivable.

Casualties Are Inevitable, Defeat Is Not

Even if ground manoeuvre forces are maintained in this way, there are challenges to ensuring that they remain effective at a relevant scale when faced with attrition. In warfighting against a peer adversary, even a professional, well-motivated and adaptable force that has been modernised and integrated along current concepts of best practice will not be able to avoid material losses and human casualties. In US doctrine, costs and risks correspond respectively to losses incurred when operations go to plan and when they do not,[44] but in either case are likely to correspond to different levels of attrition. Furthermore, successful operations have historically included tactical defeats, and successful campaigns generally include unsuccessful operations. Even the most talented and gifted commanders have experienced their share of failures, or even blunders. Prominent examples are the British campaigns in North Africa, the Far East and Europe during the Second World War, where successes at the Second Battle of El Alamein, Kohima-Imphal and Operation *Overlord* were preceded by decisive defeats in the Battle of France in 1940 and the Second Battle of Tobruk and the Battle of Singapore in 1942. These examples are extreme and characterised by

---

[43] Barzilai, *102 Days of War*, pp. 100–18.
[44] MoD, 'Joint Doctrine Note 2/19: Defence Strategic Communication: An Approach to Formulating and Executing Strategy', April 2019, p. vi-1.

mass mobilisation at a scale and for a type of warfare almost unthinkable in the post-nuclear era. However, the evidence from routine collective training is that battlegroups at British Army Training Unit Suffield and US Army brigades at the National Training Center at Fort Irwin continue to experience both failure and success as a natural consequence of facing challenging circumstances, learning from both.[45] Meanwhile, current thinking on high-end warfighting is marked by concerns about an inability for forces to recover from attrition. Even successful integration of traditional and new capabilities cannot be guaranteed to prevent attrition from rendering a force combat ineffective. In the UK, it may be argued that an obsession with manoeuvre as opposed to attritional and positional approaches to warfare often masks the fact that attrition is a constant, with the variable being the rate at which it is experienced.

The US doctrines of Multi-Domain Operations (MDO) and Joint All-Domain Operations (JADO) exemplify the current response to this conundrum.[46] In theory, the winner of the first campaign will be the overall victor, if it can thereby establish decisive overmatch against any party to the conflict whose first echelon and bespoke capabilities can be destroyed. By this logic, any subsequent campaign would be a one-sided butchery of whichever force had lost its first echelon, and the loser of the first campaign would therefore have little rational alternative but to accept an unfavourable negotiated settlement or terms. Sceptics of this hypothesis have accused MDO and JADO of falling victim to short war thinking.[47] Yet, if MDO and JADO do not fully articulate a solution to fighting a prospective high-end war, the concepts do succeed in persuasively highlighting some of its complex challenges. In particular, the potency of precision fires and other modern weapons systems leaves little margin for error for all but the largest global economies when high-end capabilities are few in number, prohibitively expensive and difficult to generate or regenerate quickly.

The lack of margin for error in such a high-end war incentivises an exclusive focus on the first echelon. The choice between an uncompetitive second echelon or no second echelon, when resources

---

[45] Alex Mills, 'Training and Mission Failure at BATUS', *British Army Review* (Vol. 174, Winter 2019), pp. 91–93.

[46] Colin Clark, 'Gen. Hyten on the New American Way of War: All-Domain Operations', *Breaking Defense*, 18 February 2020, <https://breakingdefense.com/2020/02/gen-hyten-on-the-new-american-way-of-war-all-domain-operations/>, accessed 21 March 2021.

[47] Christopher Parker, 'Rushing to Defeat: The Strategic Flaw in Contemporary U.S. Army Thinking', *Strategy Bridge*, 6 July 2020, <https://thestrategybridge.org/the-bridge/2020/7/6/rushing-to-defeat-the-strategic-flaw-in-contemporary-us-army-thinking>, accessed 29 October 2021.

expended on the former could be used to bolster the first echelon, is an obvious one – the first echelon should be the priority. However, in a 'ways, ends, means' construct where weak links between the three elements translate into increased risk, and with the theory–praxis gap and evidenced offence–defence balance having shifted, modern military forces require a first echelon that can operate and reconstitute itself at sufficient scale to survive greater-than-expected attrition. This is doable but requires both additional capacity in the combat arms and a depth of replacement or recovery, repair and refit enablers for the material element of the force. Alternatively, militaries may choose to maintain a second echelon that is capable enough to be competitive against the predicted adversary, which in practical terms will require a similar level and type of investment. Either way, the result should be that the military can fight a second campaign even after an unsuccessful first, in order to remain credible and not burdened with an unacceptable level of risk in the event of conflict. This is a high level of investment incompatible with many recent examples of 'doing more with less'. However, it may be necessary for any military wishing to deter or compete with capable adversaries. Policymakers may be forced to accept that defence after the Cold War is not destined to remain cheap.

Resilience and absorbing setbacks are not only relevant in the case of high-end warfighting at scale. A military may also need to respond to crises or otherwise operate in low-intensity conflicts or stabilisation operations that are deemed to be in the national interest. Yet, a force without sufficient mass will find itself losing its deterrent capability if it were called on to conduct a secondary campaign. Afghanistan and Iraq provide such examples for the UK, where medium-scale counterinsurgency deployments contributed to a dire lack of modernisation as conventional components of the force were neglected. A force that retains sufficient scale to maintain readiness even while supporting other commitments most importantly keeps the ability to impose or enforce thresholds on adversaries. It can also absorb setbacks in the event of conflict, afford to take more risk and potentially act in a more unpredictable manner without suboptimal battlefield outcomes equating to overall strategic defeat. Critically, it should be possible to compete with adversaries whilst concurrently retaining the necessary readiness to maintain credible conventional deterrence.

While doing more with less is an attractive proposition, it has too often been approached with an inadequate understanding of where efficiencies and force multipliers can be found, and where scale and depth cannot be replaced. Financial constraints cannot be ignored, but modern Western militaries need to reassess their discourse surrounding investment and scale to ensure that their policies match ambition with resource and can credibly achieve their stated aims. Emerging technological changes for high-end warfighting have tipped the offence–defence balance in favour of offensive capabilities, incentivising short and fast concepts of operations to knock out adversaries early. However, lean and efficient forces will find themselves vulnerable to excessive attrition and being rendered combat ineffective if they do not operate their first echelon at a larger scale than they do at present. These must include sufficient frontline ground combat forces, as platforms, systems and formations dedicated to delivering long-range kinetic and non-kinetic effects are ultimately only enablers of more traditional forces which can take, hold and dominate ground and deliver security in the areas of the battlespace they control.

IV. SWARMING MUNITIONS, UAVs AND THE MYTH OF CHEAP MASS

JUSTIN BRONK

Swarming munitions and cheap 'attritable' UAVs are two of the most common features of PowerPoint slides and forecasting documents dealing with the future battlefield.[1] Alongside the ubiquitous lightning bolts representing seamless connectivity, these highly automated assets are pictured sweeping across future skies in large numbers, rolling back the fog of war, conducting stand-in jamming and striking key targets with pinpoint accuracy.[2] It is small wonder that this vision is extremely attractive to many policymakers. In the UK, both the Chief of the Air Staff and Chief of the Defence Staff recently outlined a vision where such capabilities might provide up to 80% of the RAF's combat air mass by the 2030s.[3]

Swarming munitions are designed to be used in large numbers simultaneously, and to coordinate their actions as a group to improve overall efficiency. Attritable, reusable UAVs are an emerging class of UAV designed for a limited operational lifespan, able to carry modular sensor

---

[1] For example, Ministry of Defence, 'Joint Concept Note 1/17: Future Force Concept', July 2017, p. 43; Frank Fresconi and Scott Schoenfeld, 'ARL Experts Are on Target to Find Solutions for the Future Battlespace', US Army, 9 February 2018, <https://www.army.mil/article/200409/arl_experts_are_on_target_to_find_solutions_for_the_future_battlespace>, accessed 16 February 2021; Valerie Insinna, 'These Are the Five Areas Where the Air Force Wants to See an Explosion of Technology', *Defense News*, 17 April 2019.
[2] T X Hammes, 'Expeditionary Operations in the Fourth Industrial Revolution', *MCU Journal* (Vol. 8, No. 1, July 2017), pp. 82–103.
[3] Harry Lye, 'Future RAF Will Mix Crewed Fighters, UAVs and Swarming Drones: CDS', *Airforce Technology*, last updated 14 April 2021, <https://www.airforce-technology.com/features/future-raf-will-mix-crewed-fighters-uavs-and-swarming-drones-cds/>, accessed 9 August 2021.

and/or weapon payloads, for significantly lower acquisition and operating costs than traditional combat aircraft.[4] They represent a distinct class of weapon system from swarming munitions. However, both are united by a common appeal for military planners and policymakers; the promise of much-needed combat mass being generated cheaply through technology. In recent years, this has led to a tendency to view traditional platforms such as fast jets and cruise missiles as 'sunset capabilities' in the face of an imminent era of cheap, attritable multipurpose UAVs and swarming munitions.[5] There are certainly a range of promising trials underway which support the belief that attritable, reusable UAVs and munitions with swarming capabilities will be an important part of the future of aerial warfare. However, there are significant limitations around the potential capabilities of both classes of weapons system that need to be understood.

'Swarming' as a descriptive term for munitions denotes a specific set of capabilities which differentiate such weapons from traditional munitions which can also potentially be used in large numbers. Swarming munitions must have the datalinks, software and processing power to automatically coordinate their actions as a group in flight. They must also be equipped with sensors to build up a useful independent picture of the environment around them, so that they can coordinate their collective behaviour in response to contextual prompts. This sort of capability has been operational in limited forms since the late Cold War, especially in the realm of anti-ship missiles. The Russian SS-N-19 'Shipwreck' missile entered service in 1983 and was designed to be launched in salvos of four to eight missiles which would coordinate their actions in flight; one missile in the group would climb to higher altitudes to provide real-time radar data on targets to the others, which remained at a very low level to avoid interception.[6] If the higher-flying missile was destroyed, another

---

[4] Mark Gunzinger and Lukas Autenried, 'Understanding the Promise of Skyborg and Low-Cost Attritable Unmanned Aerial Vehicles', *Mitchell Institute Policy Paper* (Vol. 24, September 2020).

[5] Defence Synergia, 'CDS Speech to IISS 31 March 2021 – Integrated Review', 6 April 2021, <https://www.defencesynergia.co.uk/cds-speech-to-iiss-31-march-2021-integrated-review/>, accessed 29 October 2021. See also General Sir Nick Carter's discussion of 'sunset capabilities' in HM Government and Nick Carter, 'Chief of the Defence Staff, General Sir Nick Carter's Annual RUSI Speech', 5 December 2019, <https://www.gov.uk/government/speeches/chief-of-the-defence-staff-general-sir-nick-carters-annual-rusi-speech>, accessed 29 October 2021.

[6] Missile Defense Advocacy Alliance, 'P-700 Granit/SS-N-19 "Shipwreck"', 28 June 2018, <https://missiledefenseadvocacy.org/missile-threat-and-proliferation/todays-missile-threat/russia/p-700-granit-ss-n-19-shipwreck/>, accessed 5 March 2021.

from the remaining group would climb to take over the guidance and coordination role. Since the intended target – US Navy carrier battlegroups – would generally be attacked in open waters and presented a distinctive radar signature, the limited processing power and sensor capacity of Soviet missiles at the time was sufficient to enable this swarming behaviour.

For modern swarming munitions, however, the primary targets are likely to be hostile radars, missile launchers and armoured vehicles. These will generally be hidden in complex terrain, mobile, well camouflaged and often protected by decoys and point defence systems. This dramatically increases the complexity of the cooperative detection, classification and prioritisation task which modern swarming munitions must be able to undertake in flight. Nonetheless, a combination of compact, high-resolution sensors and increasingly powerful microprocessors is giving a new generation of weapons the capability to overcome many of these challenges. The US Air Force is currently testing a range of adapted cruise missiles and glide bombs to demonstrate swarming behaviours under the umbrella of its Golden Horde programme.[7] In Europe, MBDA plans to incorporate certain swarming capabilities into their SPEAR 3/EW family of munitions for the UK Ministry of Defence (MoD).[8] The key target set for both is likely to be hostile integrated air defence system (IADS) components including command and control nodes, mobile radars and high-threat surface-to-air missile (SAM) launchers.

In terms of attritable UAVs, the US Air Force recently launched its Skyborg autonomous UAV 'pilot' development programme, which commenced flight testing fitted to UAVs manufactured by various companies in summer 2021.[9] The UK's MoD has funded the development of its new Lightweight Affordable Novel Combat Aircraft (LANCA) technology demonstrator, with flight testing due to commence in 2023.[10] The Royal Australian Air Force also has six Boeing Loyal Wingman

---

[7] Thomas Newdick, 'Golden Horde Swarming Munitions Program Back On Target After Second Round of Tests', *The Warzone*, 3 March 2021.

[8] Author correspondence with MBDA subject matter expert, 8 March 2021.

[9] Valerie Insinna, 'Skyborg Makes its Second Flight, This Time Autonomously Piloting General Atomics' Avenger Drone', *Defense News*, 30 June 2021. See also Daryl Mayer, 'AFLCMC Awards Contract for Skyborg Prototypes', US Air Force, 10 December 2020, <https://www.af.mil/News/Article-Display/Article/2440755/aflcmc-awards-contract-for-skyborg-prototypes/>, accessed 18 February 2021.

[10] Craig Hoyle, 'Spirit Team to Fly LANCA Loyal Wingman Demonstrator for UK', *Flight Global*, 26 January 2021, <https://www.flightglobal.com/defence/spirit-team-to-fly-lanca-loyal-wingman-demonstrator-for-uk/142126.article>, accessed 3 March 2021.

prototype unmanned combat aerial vehicles (UCAVs) on order, with the first having made its maiden flight in March 2021.[11] The purpose of these loyal wingman-type programmes is to explore the tactical opportunities offered by teaming attritable, reusable UCAVs with fast jets, and to guide the rapid acquisition of UCAV capabilities to augment existing fleets. As such, the UCAVs will need to be able to carry fuel, sensors and weapons at speeds and over comparable distances to those jets.

Anything that flies must balance aerodynamic, size and performance attributes which directly impact cost, capabilities and potential uses. The range that any munition or aircraft can fly will depend on the fuel efficiency of its engine, the amount of fuel that can be carried, desired cruise speed and altitude, and the lift/drag ratio of its aerodynamic configuration. The larger an airframe is, the more weapons and sensors can be carried as a given proportion of its usable weight and internal space. This is why combat aircraft have become progressively larger since their inception, a trend enabled by more powerful and efficient engines. A modern fast jet must carry a wide array of multirole weapons, a heavy and complex radar and other sensors, which all require power, cooling capacity and computer hardware to control them and process the information they generate, as well as sufficient fuel to provide the required performance over extended ranges. Providing suitable range and performance while accommodating these payload and space requirements is the primary reason why a modern fighter jet such as the RAF's Typhoon FGR4 weighs around 21 tonnes when loaded, despite the extensive use of composite materials to keep weight to a minimum.[12] Only a relatively small proportion of the total aircraft weight and size is directly attributable to the need to accommodate crew, a cockpit with instrumentation and life support systems, and ejection seats. 'Loyal wingman' or less explicitly tethered, more autonomous attritable UAVs can avoid the weight and cost penalties associated with a crew. However, they will still be bound by the same physics-based trade-offs that have so far led to consistent weight and size growth in comparable combat aircraft types with each successive generation.

Put simply, if 'attritable' UAVs are to be capable of similar performance to fast jets and are to carry similar weapons and sensors, they will either have to be comparable in size and weight, have a much

---

[11] Greg Waldron, 'Boeing Australia's "Loyal Wingman" Conducts Maiden Sortie', *Flight Global*, 2 March 2021, <https://www.flightglobal.com/defence/boeing-australias-loyal-wingman-conducts-maiden-sortie/142685.article>, accessed 3 March 2021.

[12] Royal Air Force, 'About the Typhoon FGR4', <https://www.raf.mod.uk/aircraft/typhoon-fgr4/>, accessed 19 September 2021.

shorter range, or a much smaller useful payload. The weight of any aircraft is closely linked with operating costs, since in conjunction with the aerodynamic configuration the weight dictates the amount of power needed from the engines in order to produce sufficient lift at various speeds. Heavier airframes need more powerful engines to fly at any given speed, which generally consume more fuel. This results in either a shorter range or the need to carry more fuel, which in turn further increases the required size and weight of the aircraft. As such, the lighter a UAV can be made, the more likely it is to be sufficient to allow affordable operations at significant scale. However, this will prevent it from carrying large payloads, and also limit either range or performance.

Sacrificing fast jet-class performance in favour of slower cruising speeds would allow range to be increased relative to airframe size and weight. The slower something is required to fly, the less aerodynamic drag it must overcome and so the smaller and more fuel efficient its engine can be. Thus, if desired cruise speed (and acceleration characteristics) can be kept low, then range for attritable UAVs can be significantly extended for a given size and fuel capacity. However, this would mean that such attritable UAVs would need to operate in loose coordination with, rather than in mixed formations alongside, fast jets in 'loyal wingman'-type roles. A similar dynamic is true for swarming munitions development – for a given size, weight and cost of munition, speed can be increased at the expense of range or vice versa, but militaries cannot have both in the same design. If something must travel fast and over long distances, it will have to carry a lot of fuel to feed a powerful engine, and thus be larger, heavier and more expensive.

Stealth properties further shift this equation towards high acquisition and operating costs. The airframe shapes which are compatible with low observability to radars operating in the X and Ku bands are generally less aerodynamically efficient than more traditional aircraft shapes. This means they must be larger to generate the same amount of lift. Furthermore, aircraft which rely on low observability to survive against hostile forces must carry all their fuel and weapons internally, since traditional stores on external pylons greatly increase radar cross section. Therefore, stealth aircraft must be larger and more complex than traditional equivalents due to the need to accommodate internal weapons bays and all the necessary fuel within the airframe itself. The stealth coatings which typically cover such aircraft are also more difficult and expensive to manufacture than traditional aircraft skins and must be maintained to a high level of finish to remain effective, which increases the costs of maintenance and storage. The widespread adoption of stealth aircraft also affects the requirements for both swarming and more traditional munitions. Since munitions delivered by stealth aircraft must be carried internally to avoid compromising radar signature, they are subject to more restrictive size, weight and shape constraints than munitions which are only intended for external carriage. There is also pressure to develop smaller, shorter-ranged munitions for stealth aircraft since the latter can get closer to threats, which reduces the value of additional weapon standoff range, while being constrained in terms of weapons carried per aircraft.

In summary, there are a series of relatively inescapable trade-offs which will dictate how and where attritable, reusable UAVs and swarming munitions will augment and potentially replace more traditional aircraft and weapons systems. Cost is arguably the key element for both, since the concepts of 'attritable' UAVs and munitions designed to be expended in large numbers both rely on such weapons being bought in sufficient numbers to be expended sustainably in conflict. Furthermore, the same systems that are attritable and/or fieldable en masse for the US or China are unlikely to be viewed in the same way by smaller allies or export customers.

There are three broad mission sets for which 'attritable' UAVs are being developed. The first is the comparatively straightforward 'loyal wingman' role wherein UCAVs with similar performance to piloted fast jets operate alongside the latter in cooperative tactical units. The second is more independent UCAV operations with airframes in a similar weight class as fast jets. These would conduct standalone sorties in support of piloted assets, especially in the ISTAR and suppression/destruction of enemy air defences (SEAD/DEAD) roles. The third mission set is for UAVs in a similar weight class to traditional cruise missiles to conduct primarily ISTAR and electronic warfare missions. These would be released and potentially recovered by carrier aircraft or launched by surface units via booster rockets and recovered by parachute.

For loyal wingman-type UCAVs, the most significant cost and size drivers are likely to be the requirement for sufficient performance and range characteristics to maximise tactical interoperability with the piloted fast jets which they will fly alongside. If intended to operate alongside stealth aircraft without giving away the latter by their presence, such loyal wingman UCAVs will also need to carry advanced sensors and munitions internally within a low-observable airframe. These performance, range and payload requirements mean that combat-capable loyal wingman-type UCAVs are likely to end up being at least a similar size and weight to advanced jet training aircraft such as the BAE Systems Hawk or Boeing/Saab T-7A.

They will also have to have significant capacity for in-flight autonomy, including the ability to defend themselves and make tactical decisions to maintain coordination with potentially unpredictable human formation members under rapidly changing circumstances.[13] The loyal wingman concept allows human fighter pilots to remain 'on the loop' in terms of giving tactical directions to their accompanying UCAVs through line-of-sight datalinks. However, pilots do not have limitless mental capacity, and can get task saturated simply managing their own aircraft, weapons and communications, and trying to maintain situational awareness in combat.[14] As such, a loyal wingman-type UCAV must be fully combat capable even when only receiving relatively limited real-time directions and permissions by their human formation mates. Unlike many current-generation remotely piloted UAVs, loyal wingman-type UCAVs must also be capable of autonomous launch and recovery capabilities in any weather conditions that their piloted fighter stablemates would be expected to operate in, since they will provide a core part of the 'system of systems' which will supposedly enable those piloted fast jets to remain combat effective against future threats. These requirements will drive significant software complexity, processing power and sensor requirements, and imply the need for serious national debates and governmental policy decisions on whether the required levels of lethal autonomous capabilities are acceptable.

Projects such as Skyborg, LANCA and Loyal Wingman are producing advanced prototypes for a whole new class of combat aircraft, but the costs, airframe weight and complexity levels for these proof-of-concept projects are significantly lower than those which a full-scale frontline loyal wingman UCAV procurement will imply. This class of UAVs certainly offers the potential promise of relatively cheap combat aircraft that will unlock a slew of potentially very useful tactics for use alongside piloted fast jets. However, it remains to be seen whether they can be developed to be genuinely combat-capable and dependable frontline aircraft for high-intensity conflict scenarios at a cost that allows them to be procured in sufficient numbers to be seen as 'attritable' by most air forces. They are also unlikely to be particularly attractive to air forces for operations in permissive airspace, where remotely piloted, long-endurance systems such as the ubiquitous MQ-9 Reaper series will remain more efficient.

---

[13] There is a similar phenomenon at play with loyal wingman-type UCAV autonomy requirements, as has been found by developers of driverless cars. Safe and predictable interactions are easier to programme between multiple automated vehicles than a mix of automated and human-controlled ones. See, for example, Rodney Brooks, 'The Big Problem With Self-Driving Cars Is People', *IEEE Spectrum*, 27 July 2017.

[14] *Fast Jet Performance*, 'SAM Dodging Over the Nevada Desert – Why Low-Level Flying Is Still Necessary', 2015, <https://www.fastjetperformance.com/blog/sam-dodging-over-the-nevada-desert-why-low-level-flying-is-still-necessary>, accessed 9 March 2021.

To avoid the programming complexity, datalink reliance and performance requirements inherent in loyal wingman-type UCAV development, many countries have developed UCAV demonstrators which appear optimised for more independent concepts of operations. This category of UCAVs generally consists of flying wing- or cranked kite-type airframes with buried turbofan engines optimised for minimal broadband radar and infrared signature, and subsonic flight.[15] This makes them ideal for efficient long-range penetrating flights into heavily contested airspace, with obvious potential uses in terms of ISTAR, SEAD/DEAD and strikes against high-value targets. However, they are less optimised for close tactical cooperation with piloted fast jets, since they have lower performance, much more restricted agility and are designed for survivability through remaining undetected. The X-45 and X-47 programmes in the US, Taranis in the UK, nEUROn in France and GJ-11 in China are all examples of technology demonstrators for this class of UCAV.[16] The software and processing capabilities required for an operational UCAV of this type in the standalone ISTAR, SEAD/DEAD or even strike role against pre-planned targets are in many ways less advanced than those required for loyal wingman operations. This is because they do not have to optimise their in-flight behaviour for effective tactical cooperation and safe collision avoidance with a potentially unpredictable human pilot. Also, many of the key targets that they might be tasked with finding and attacking in a high-intensity conflict (such as IADS components) have very distinctive radar and IR signatures which helps to reduce assurance requirements for rules of engagement.

The lethal autonomy implications, however, are even starker than in the case of loyal wingman-type operations since standalone UCAVs offer the greatest potential advantages in mission sets which involve penetrating heavily defended airspace without direct support. In such airspace, long-range datalinks or satellite communications are unlikely to be reliable or suitable for use for real-time control of weapons release decisions. This is unlikely to prevent major powers such as the US, China and Russia from developing penetrating UCAVs, meaning that the

---

[15] See, for example, Northrop Grumman, 'X-47B UCAS', <https://www.northropgrumman.com/what-we-do/air/x-47b-ucas/>, accessed 9 March 2021; BAE Systems, 'Taranis', <https://www.baesystems.com/en/product/taranis>, accessed 9 March 2021.

[16] For a detailed discussion of each, see Justin Bronk, *The Future of NATO Airpower: How Are Future Capability Plans Within the Alliance Diverging and How Can Interoperability be Maintained?*, Whitehall Paper 94 (London: Taylor and Francis, 2020).

question for medium powers with aspirations for penetrating ISTAR, SEAD/ DEAD and strike capabilities is *when* rather than *if* they take the ethical and political plunge. With the requisite high levels of in-flight autonomy, standalone UCAVs can be designed with a low airframe fatigue life and operated sparingly in terms of hours flown to retain combat proficiency of each squadron compared to piloted aircraft. This means total through-life costs of such systems are likely to be significantly lower than those of piloted equivalents. However, the very-low-observable airframe, autonomy and sensors required to operate in the penetrating role in defended airspace over comparable ranges to piloted combat aircraft once again imply an acquisition cost which is lower but still comparable to those of fast jets. Since they offer clear benefits for high-intensity mission sets, penetrating low-observable UCAVs are likely to be a major feature of high-end state air power in the coming decades. However, although one of their attractive qualities is the lack of risk to human crew, they are likely to be too expensive to be considered truly attritable for any but the largest of air forces.

The third important class of attritable UAVs being trialled at the time of writing are those in or below the size and weight class of cruise missiles. These UAVs blur the line between autonomous aircraft and advanced, potentially reusable munitions. The upper end of this category involves compact, subsonic airframes with a small jet engine, fold-out wings and a modular sensor and/or electronic warfare payload capacity. They are designed to be either launched from other aircraft or by using a booster rocket from the ground or a maritime platform, and then potentially recovered by parachute or in-flight docking with another aircraft after they have fulfilled their missions. DARPA's 'Gremlins' programme is perhaps the best known, but there are likely experiments being undertaken by many states to develop similar concepts.[17] The attraction of such systems is that they can potentially be reused, adding efficiency over single-use loitering munitions or stand-in jammer equivalents, while also having an acquisition cost measured in hundreds of thousands or low millions of dollars as opposed to tens of millions for loyal wingmen or standalone UCAVs. The limitations compared to those systems concern payload, mission flexibility and a likely cost which is still comparable to high-end cruise missiles, which most medium and small air forces already struggle to field in sufficient numbers. Ultimately, these UAVs offer various cost advantages over single-use munitions in certain contexts and can allow the application of novel swarming tactics to problem sets such

---

[17] Paul J Calhoun, 'Gremlins', *DARPA*, <https://www.darpa.mil/program/ gremlins>, accessed 8 March 2021.

as SEAD/DEAD or ISTAR within defended airspace to provide eyes forward for an incoming strike package. However, due to the additional complexity implied by their reusable nature and modular multipurpose payload options, they will tend to remain more expensive than single-use swarming and conventional standoff munitions. Some projects such as the RAF's Alvina programme are examining significantly smaller and cheaper systems, but these will be very limited in range compared to larger UAVs and piloted combat aircraft, and/or sacrifice responsiveness by having a slow cruising speed.[18] With short-range systems, potentially vulnerable delivery platforms will have to get dangerously close to threats whilst carrying a large payload of still relatively expensive 'attritable' UAVs.

The trade-offs inherent in survivability, range, payload and cost have already forced a split in traditional standoff missile development between high-supersonic/hypersonic missiles or more ambitious stealth features on subsonic ones. It is almost impossible to have both hypersonic performance and a low infrared and radar signature due to the aerodynamic shapes and propulsion technologies required for each being highly specialised. Both techniques significantly increase the cost per weapon, which in turn has contributed to a demand from militaries for smaller, cheaper and 'smarter' swarming weapons. The idea is that by coordinating their actions and providing too many threats for IADS to intercept simultaneously, swarming weapons can offer the required levels of lethality against improving defences without the cost associated with comparable numbers of cruise missiles, decoys and jammers.

The most basic and achievable type of swarming logic requires munitions to be equipped with datalinks so that each can update the others about their position, chosen target and status, and then use pre-programmed algorithms to ensure targets are not duplicated. More advanced systems such as modern anti-ship missiles will also coordinate their flight paths and final approaches to target vessels to minimise the effectiveness of the defences on whatever ships they are attacking. One way of doing this, for example, is to have missiles make their run-ins to the target ship from the bow and the stern quarter simultaneously. Smaller munitions can also target weak or critical areas of hard targets such as reinforced structures or ships, so that a number of small warheads can collectively cause disproportionate damage. Moving further up the scale of autonomous swarm capabilities, weapons like those

---

[18] Craig Hoyle, 'RAF Chief Reveals Combat Cloud, Swarming Drone Advances', *Flight Global*, 15 July 2021, <https://www.flightglobal.com/defence/raf-chief-reveals-combat-cloud-swarming-drone-advances/144604.article>, accessed 9 August 2021.

being trialled in the Golden Horde initiative search for, classify and collectively prioritise targets as a group, reacting dynamically to unexpected threats or changes in the environment.[19]

The degree of autonomous, cooperative capabilities which munitions are developed with will depend on the anticipated breadth and complexity of potential mission sets, and the cost, weight and range requirements to be met. A small munition such as SPEAR 3 or the GBU-53/B StormBreaker, for example, cannot carry a sufficiently powerful warhead to breach deeply buried or heavily fortified targets such as bunkers. Furthermore, if a weapon is to be capable of multiple mission sets in different weather conditions, this will increase the requirements for multispectral, high-resolution sensors such as millimetric radar, infrared imaging or light detection and ranging (LIDAR). However, multispectral sensors also significantly increase weapon cost and require additional space, batteries and coolant. The latter compete with fuel, warhead options and the propulsion system for space and weight. Once again, the longer-ranged and more responsive, flexible and precise a swarming weapon is required to be, the larger and more expensive it becomes.

There is also the issue of launch platform capacity and risk tolerance. The more survivable a launch platform is, the closer it can get to threat systems and target areas to deploy swarming munitions. Getting closer also allows launch platforms more opportunities to use their onboard sensors to give the munitions the best situational awareness picture possible at launch. Stealth fighters are an obvious candidate but are limited in terms of munitions capacity by having to carry them internally to maintain a low radar signature. More traditional fighters such as the Typhoon or F-15E can carry more munitions but must stand off further from hostile threats for a given level of risk. Large bombers or transport aircraft can potentially carry large numbers of munitions but are easy to detect and have very limited self-defence capabilities if fired upon. Therefore, there is an inverse relationship between how close a launch platform can get to defended targets and the number of munitions each launch platform can carry. This is further complicated by the fact that having to launch from longer distances from the target area means that the munitions must be larger or much slower to achieve the required range. A munition in the 100-kg class with an efficient mini-turbojet propulsion system might be able to offer standoff ranges of around 150 km with a medium–high altitude release. However, this is still well

---

[19] Garrett Reim, 'Golden Horde Collaborative Bombs Hit Four Targets Simultaneously in Test', *Flight Global*, 26 February 2021, <https://www.flightglobal.com/fixed-wing/golden-horde-collaborative-bombs-hit-four-targets-simultaneously-in-test/142655.article>, accessed 1 April 2021.

within threat range for legacy fighters, let alone transport aircraft, against a modern peer IADS. Therefore, swarming munitions intended to be launched in numbers beyond those which can be carried by the limited number of penetrating assets available will need to be larger and heavier, thus making them more expensive.

The addition of swarming capabilities promises significant new tactical capabilities to a range of existing large and small munitions classes. These will increase their ability to overcome point defence systems and allow the integration of organic stand-in jamming capabilities to a salvo of kinetic munitions. More sophisticated examples will also allow cruise missiles and glide bombs to search areas for suspected but as-yet-undetected threats, where rules of engagement permit. However, these are capability enhancements which will come with a high cost per weapon fired, even compared to current-generation standoff weapons. They will also not break the basic relationship between desired standoff range, cruise speed and weapon size/cost. Thus, any assumption that swarming munitions will solve current mass shortcomings in terms of munitions stocks, and the lack of sufficient survivable launch platforms to get them within range of defended targets, is sadly misplaced.

In conclusion, swarming munitions and sophisticated UAVs promise a range of opportunities to increase the efficiency of projecting power in the air domain. However, unit costs will remain high, especially if the systems in question are intended to be sufficiently flexible to replace traditional alternatives across a range of operational scenarios, rather than provide incremental efficiency gains through a supplementary approach. Even following the latter approach, the required attributes for most roles mean that, for countries other than the US and China, swarming munitions and UAVs/UCAVs will only be affordable in small quantities.

V. THE LIGHTS MAY GO OUT, BUT THE BAND PLAYS ON

PETER ROBERTS

The likelihood of success in achieving national foreign policy goals against a competitor may be drawn from a set of characteristics determined by a view on the threat being posed. In military circles, this has historically been equated to the idea that a threat is equal to the military capability of a state plus their intent to carry out activities against you.[1] This methodology was rapidly adopted by business seeking 'competitive advantage' and became *lingua franca* across the public, private, commercial and military domains before 2005.[2] In military education, where future senior officers and their civil service counterparts are groomed for high office, such basic calculations have become key indicators of military prowess – applied as much as an assessment of one's own ability to enact foreign policies as it has to other belligerents.[3]

In national security terms, and increasingly since 1945, this same calculation has gradually been refined by various states to one that places equal emphasis on military capability and will to fight. Indeed, given the history of Western states, the idea of a will to fight (the intent part of the calculation) has become fixed – first by the ideological position of the Soviet Union as a long-term adversary, later by the idea of terrorism as a singular amorphous entity, and most recently by various insurgent groups

---

[1] J David Singer, 'Threat-Perception and the Armament-Tension Dilemma', *Journal of Conflict Resolution* (Vol. 2, No. 1, 1958), pp. 90–105.
[2] Gary Hamel and C K Prahalad, 'Strategic Intent', *Harvard Business Review* (July–August 2005).
[3] Peter Roberts and Heather Venable, 'Episode 62: Heather Venable: Gen-Z – The Best Tacticians in History?', RUSI Western Way of War podcast, 2 September 2021, <https://www.rusi.org/podcasts/western-way-of-war/episode-62-heather-venable-gen-z-best-tacticians-history>, accessed 28 October 2021.

such as Al-Qa'ida, the Islamic State and the Taliban. In this, political and military leaders seem to have developed an idea that their state's will to fight is a fixed variable that has been in a state of slow decline.[4] As such, the chief variable in determining success on the battlefield has often been equated to military spending alone and the associated military equipment that is delivered and in service. This is not only a Western issue – for example, Russia's doctrine of a 'correlation of forces' considers military capabilities as lined up against each other in a cold calculation that the greater capability will win on the battlefield.[5] Since the concept of military 'mass' is thought to be unaffordable in contemporary Western military circles, technology has become the crutch with which military and (later) political leaders have sought to counterbalance the equation in their favour.[6]

However, this chapter will evidence that states have proven highly resilient and adaptable in response to shocks over time. Unless they are physically occupied, states are usually able to reconstitute and conduct prolonged operations. In time, these states have demonstrated an ability to withstand continuous and considerable pressure from foreign actors. Furthermore, casualty tolerance is not a fixed variable but is highly context dependent. In any major conflict, casualty tolerance often increases drastically but is also variable over time: publics have shown themselves to be much more tolerant to casualties than their political or military leaders, depending on the cause. This idea of resetting the narrative on a national will to fight in political and military circles is vital in order to embed a sound conceptual understanding of our ability to wage war in the future.

This chapter aims to refine and redefine this calculation to one that places a much higher emphasis on the national will to fight. It outlines the idea that a nation's military capability multiplied by the national will to fight must be greater than that of an enemy to prevail in a conflict. Importantly, the idea of the national will to fight – linked to a political will, which is itself derived from popular support for an intervention – has two associated factors. First, that the national will to fight multiplies military capability and is not a simple addition, and second that the will

---

[4] Michael J McNerney et al., *National Will to Fight: Why Some States Keep Fighting and Others Don't* (Santa Monica, CA: RAND, 2018).

[5] Julien Lider, 'The Correlation of World Forces: The Soviet Concept', *Journal of Peace Research* (Vol. 17, No. 2, 1980), pp. 151–71.

[6] Ministry of Defence (MoD) and Nick Carter, 'Chief of Defence Staff Speech RUSI Annual Lecture', 17 December 2020, <https://www.gov.uk/government/speeches/chief-of-defence-staff-at-rusi-annual-lecture>, accessed 23 August 2021.

to fight a specific campaign is not deterministic but contextual and highly variable.

The chapter concludes that when evaluating the risks in any potential conflict, and the likely outcomes, policymakers should pay careful attention to the relative strengths of their second echelon forces and not simply rely on those asked to fight the first battle. Campaigns are rarely short, and there are fewer examples of wars that are determined by the initial engagements. Instead, political and military leaders might do better to understand their models for fighting over longer periods of time. This requires a different engagement with the public in ensuring long-term support for key causes: the current self-imposed limitations of policymakers assume public intolerance for casualties that deters the creation of realistic policies, but are also a dubious set of biases that bears little basis to the lived experience of war.

In its simplest form, the political will to fight is a reflection of the attitudes of the general public rather than an informed, nuanced and long-term view to any conflict, intervention or overseas activity. Such attitudes change in relation to perceptions of the amount at stake in any engagement, the benefits that might be derived from a military activity, and the narrative being used to explain and illustrate the campaign, the context and the conditions. Yet, in acknowledging the realities of these attitudes, which will naturally vary over the course of any campaign, the public and their politicians will be both fickle and aware of the 'sunk costs' of longer military adventures.[7]

Resilient Societies

Western societies have proven themselves remarkably resilient to systemic shocks. Indeed, in many cases they have exceeded expectations in terms of their ability to adapt to disruption. Consider, for example, the case of Britain over the course of the Napoleonic and world wars. The country's dependence on food imports and losses to blockades were at their greatest during the First and Second World Wars, yet suffering from food shortages during the Napoleonic era exceeded that of the two later conflicts. The modern state's ability to reorganise, stockpile and substitute lost imports largely mitigated the effects of a strategy of punishment,

---

[7] 'Sunk costs' refer to the amount of resource that has been already expended and cannot be recovered. In national security terms, there is an associated feeling of 'how can we give this up after all we've sacrificed?'. See Carl Forsling, 'Understanding the Sacrifice and Sunk Cost of the War in Afghanistan', *Task and Purpose*, 11 March 2021, <https://taskandpurpose.com/analysis/afghanistan-sunk-costs-withdrawal/>, accessed 29 October 2021.

even when a critical resource was threatened.[8] The Royal Navy and British-flagged merchant fleets were essential in providing protection to convoys and transporting supplies during this period. While they sustained heavy losses, the state itself proved remarkably resilient.

Such examples worked against the logic of military strategists of the time. For example, before the Second World War, advocates of air power such as Guilio Douhet made the case that the destruction of the infrastructure that underpinned an opponent's civilian economy would render direct clashes with its forces unnecessary.[9] An analogue to this notion was also voiced by figures such as Basil Liddell Hart and the Defence Secretary Leslie Hoare Belisha, both advocates of the so-called 'indirect approach' through which a combination of blockades and airpower would allow Britain to avoid a repetition of the bloody attritional battles of the First World War.[10] Their ideas were founded on an argument that societies were complex and fragile ecosystems in which the disruption of key subsystems could produce the collapse of the system as a whole.[11]

More recently, much has been made of the use of combined powers of states to undermine their adversaries at a national level. Russian operations in Georgia (2008) and Ukraine (2014) used political, military, economic and cyber levers to attack sovereign governments and achieve a degree of success in meeting their desired goals. Yet, in doing so, Russia also galvanised the attacked societies to solidify as states in response and harden their own resilience to future attacks. In Georgia, successive administrations overturned NATO advice and reverted to a conventional military, moving away from the specialised and technical one that other Western powers had been advocating. In Ukraine, a 40,000-person, crowd-funded militia demonstrated public support for a

---

[8] Mancur Olson Jr, *The Economics of the Wartime Shortage: A History of British Food Supplies in the Napoleonic War and in World Wars I And II* (London: Literary Licensing, 2012).

[9] Robert A Pape, *Bombing to Win: Air Power and Coercion in War* (Ithaca, NY: Cornell University Press, 1996), p. 60.

[10] John J Mearsheimer, *Liddell Hart and the Weight of History* (Ithaca, NY: Cornell University Press, 1988), p. 103. In fairness to Liddell Hart, he did accept the importance of attrition on the ground, but felt that this role should be outsourced to allies.

[11] This idea had a particular (though not exclusive) appeal to classical liberals who feared the impact of large standing armies on the power of the state and looked to win wars in an economical way. See, for example, Azar Gat, *A History of Military Thought: From the Enlightenment to the Cold War* (Oxford: Oxford University Press, 2001), pp. 200–50.

grinding attritional campaign that has withstood continual Russian incursions for another seven years.

China may have been learning from Russia's more obvious intrusions that have elicited national-level responses. Consistent with its doctrine of fighting 'local wars under informatized conditions',[12] China appears to view disruptive cyber attacks at the strategic level – which might lead a target population in the US to be more invested in a conflict due to the anger they produce – as being counterproductive. Chinese leaders view the wartime applications of cyber attacks in tactical and operational terms, targeting military systems and C4ISR nodes in conjunction with kinetic attacks.[13] Espionage against strategic targets has not been matched by a pattern of behaviour, suggesting that subversion and destruction will be part of a Chinese local war campaign.

There are, of course, examples of states being less resilient to the coercive use of force. In the aftermath of the terrorist bombing of trains in Spain in 2004, the Spanish government lost the election and withdrew from supporting the war in Iraq. Whilst achieving the initial aims, the attacks did not drive a wider withdrawal of support from US-led forces in the Multi-National Forces Iraq. This was a stated aim of the published memo of the Global Islamic Media, which had labelled Spain as 'the domino piece most likely to fall first'.[14]

It is impossible to categorically state that all attacks against the infrastructure of states will or will not weaken a will to fight. But neither is it possible to predict under what circumstances the same actions will or will not strengthen a state's will to fight. The broad examination of historical evidence indicates that states are more resilient than they are given credit for. There is no useful guide to which coercive tools support the outcomes an actor might desire, but it seems that there is a good deal more relevance to be placed on the interaction of actions on the societal culture that might alter the national will to fight.[15]

The metrics for measuring the resilience of a society to shocks and how they change under coercion and war, their ability to react and their flexibility to rebound, are complex and unwieldy. A clearer measure might be found in the reaction, tolerance and acceptance of casualties.

---

[12] Edmund J Burke et al., 'People's Liberation Army Operational Concepts', RAND Corporation, 2020.

[13] Jon R Lindsay, 'The Impact of China on Cybersecurity: Fiction and Friction', *International Security* (Vol. 39, No. 3, 2015), pp. 7–47.

[14] Soufan Center, 'IntelBrief: 15 Years After Madrid Tarin Bombings, What Have We Learned', 11 March 2019, <https://thesoufancenter.org/intelbrief-the-15th-anniversary-of-the-madrid-train-bombings/>, accessed 10 October 2021.

[15] Theo Farrell, 'Culture and Military Power', *Review of International Studies* (Vol. 24, No. 3, 1998), pp. 407–16.

It Is Not the Casualties but the Cause

The avoidance of unnecessary casualties is a desirable military tenet, usually framed as economy of force in the Western principles of war,[16] that overlaps all methodologies of warfare at the operational level. The desire, in these terms and at this level, is to preserve resource better than an adversary in a contest of wills that will, necessarily, be attritional to some degree. More recently, as militaries have been professionalised and become smaller, the reduction in casualties has been increasingly important to preserve scarce resource and prolong output from expensive capabilities that take considerable time to generate.[17]

There is another part of the discussion over casualties in conflict that has become an axiom: that society (and thus politicians) have no stomach for military casualties in warfare.[18] There is an argument made on this basis that militaries have less utility in 'challenging' circumstances where the risk of loss of personnel cannot be mitigated, and that in attempting to adapt to this new policy of casualty aversion, militaries have had to rely increasingly on physical barriers, technology and different ways of engaging to reduce risk. The argument that these are less effective in allowing human-to-human contact requires a balance to be struck, but one in which politics and societal risk aversion plays a crucial role.[19]

The third element of casualty aversion is associated with civilian casualties during military action. As military forces have adapted ways of fighting that minimise their own casualties (notably in the use of drones and standoff munitions to kill targets), the result has been a direct correlation to the public perceptions of increased civilian casualties. Until recently, there was little useful research into this part of understanding public support for military action. Nonetheless, recent studies have produced some useful data and polling – albeit nulling some variables

---

[16] Jan Angstrom and J J Wigen, *Contemporary Military Theory: The Dynamics of War* (New York, NY: Routledge, 2015), pp. 78–79.

[17] Peter Roberts and Tony King, 'Episode 30: Is the Era of Manoeuvre Warfare Dead?', RUSI Western Way of War podcast, 10 December 2020, <https://rusi.org/podcasts/western-way-of-war/episode-30-era-manoeuvre-warfare-dead>, accessed 29 October 2021; Peter Roberts and Mungo Melvin, 'Episode 23: Utility vs Utilisation', RUSI Western Way of War podcast, 5 November 2020, <https://rusi.org/podcasts/western-way-of-war/episode-23-utility-vs-utilisation>, accessed 29 October 2021.

[18] Niklas Schörnig and Alexander C Lembcke, 'The Vision of War Without Casualties: On the Use of Casualty Aversion in Armament Advertisements', *Journal of Conflict Resolution* (Vol. 50, No. 2, 2006), pp. 204–27.

[19] Thomas R Mockaitis, *Civil-Military Cooperation in Peace Operations: The Case of Kosovo* (Carlisle, PA: US Army War College, 2004).

that could have significant implications for their findings.[20] These three factors have resulted in a political casualty *intolerance* that has been a focus of military planning in the UK at least since 1996 and, according to some, has shaped the country's government policy about interventions, engagements and national security since 2003.[21]

The fixation on military casualty rates is a relatively recent phenomenon, linked particularly to post-Cold War military engagements, and has been variable throughout campaigns. A direct link between the nature of public outcry and interest in military interventions is seemingly linked to perceptions of the campaign itself. For example, deaths during the 2003 invasion of Iraq received much less attention than the subsequent deaths of service personnel during the long counterinsurgency/state-building campaign that followed. Similar experiences can be found during the campaign in Afghanistan: casualties in Operation *Jacana* in 2002 were expected, but during the subsequent state-building campaign (2004–17), personnel losses were met with more questions and resistance.

The same is true for US experiences, as well as other allies who engaged in these campaigns. One might draw the conclusion that the public expects the military to make sacrifices in wars they believe are worthwhile but will be less tolerant when the benefits are unclear. They also change over the course of these campaigns. This has significant implications for the military, military operations, and political decision-making in questions of defence and security.

Arguably, Western governments have always had a requirement to justify their use of military forces to the electorate. As an intrinsic part of the democratic process, how governments explained why they were waging war also involved a narrative of how they were fighting it; an explanation for the costs in 'blood and treasure' to the state and society at large.[22] The number, scale and severity of casualties (dead and injured) during conflict started to become a more prominent question in proxy conflicts during the final stages of the Cold War.[23] Indeed, it has been argued that the Vietnam War was a turning point in the US – taking

---

[20] Robert Johns and Graeme A M Davies, 'Civilian Casualties and Public Support for Military Action: Experimental Evidence', *Journal for Conflict Resolution* (Vol. 63, No. 1, 2019), pp. 251–81.

[21] Paul Cornish, 'Myth and Reality: US and UK Approaches to Casualty Aversion and Force Protection', *Defence Studies* (Vol. 3, No. 2, 2003), pp. 121–28.

[22] John E Mueller, *War, Presidents and Public Opinion* (New York, NY: John Wiley and Sons, 1973), pp. 60–69.

[23] Victor Mahieu, 'Casualty Aversion in Western Countries', Finabel, 15 May 2019, <https://finabel.org/casualty-aversion-in-western-countries/>, accessed 12 February 2021.

time to move across the Atlantic to other Western allies – and produced new attitudes towards military casualties in the Western world, stimulating the term 'Vietnam Syndrome' to indicate the 'American public's unwillingness to continue to support US foreign military efforts, particularly as casualties rise'.[24] Elsewhere in the West, timelines might have been different but thinking was shifting in the same direction: casualties in the Falklands War (1982), and how the dead were commemorated,[25] alongside the long-running counterinsurgency operations in Northern Ireland did not shape policy decisions in the UK at the time, yet casualty aversion did become a significant factor in the 1990s. Subsequently, it has taken up permanent residence in national security decision-making.[26]

It was only really after the end of the Cold War that the political and military concern over casualties as such became an issue across all Western states. The images of dead and injured service personnel in Somalia (1993) or in the Former Republic of Yugoslavia (1995) brought home the cost of conflict in a world lauded as largely devoid of enemies.[27] For some allies, it has been claimed that such has been the lack of risk appetite for casualties that it has become the single factor in determining whether interventions will take place or not.[28] Others claim that the casualty averse nature of political decision-making since the 1990s has forced militaries to embrace an entirely new Western military concept of war, one in which rapid, bloodless victories are possible when technology and overwhelming superiority can be assured.[29] Notably, this research took place before the global War on Terror following terrorist attacks in 2001. Yet, the rhetoric of risk aversion has again made its way into the public discourse by senior military leaders.[30]

Under the broad category of casualties (including those projected in 'collateral damage' estimates), one might consider classifying causes for casualty aversion under four headings: national interest; strategic

---

[24] Richard A Lacquement Jr, 'The Casualty-Aversion Myth', *Naval War College Review* (2004), p. 41.

[25] Helen Parr, *Our Boys: The Story of a Paratrooper* (London: Allen Lane, 2018).

[26] Christopher Coker, 'Post-Modern War', *RUSI Journal* (Vol. 143, No. 3, 1998), pp. 7–14.

[27] Charles A Stevenson, 'The Evolving Clinton Doctrine on the Use of Force', *Armed Forces & Society* (Vol. 22, No. 4, 1996), pp. 511–35; Jan van der Meulen and Marijke de Konink, 'Risky Missions: Dutch Public Opinion on Peacekeeping in the Balkans', in Philip Everts and Pierangelo Isernia (eds), *Public Opinion and the International Use of Force* (New York, NY and Abingdon: Routledge, 2001).

[28] T S Milburn, 'Casualties – The Crucial Factor in Modern Conflicts', *British Army Review* (Vol. 113, August 1996), pp. 78–84.

[29] Andrew P N Erdmann, 'The U.S. Presumption of Quick, Costless Wars', *Orbis* (Vol. 43, No. 3, Summer 1999), pp. 378–80.

[30] MoD and Carter, Chief of Defence Staff Speech RUSI Annual Lecture'.

calculus; internal politics; and long-term social change.[31] In doing so, one must also acknowledge the military drive for reduced casualties that has endured for millennia in both the East and the West, whether in full-spectrum, high-intensity conflict or in more coercive and less kinetic campaigns. Both Sun Tzu ('Supreme excellence consists of breaking the enemy's resistance without fighting') and Machiavelli ('Never attempt to win by force what can be won by deception') advocated for approaches that maximised gains and minimised casualties.[32] Warfare can never be regarded as an efficient use of national resource, and the loss and/or waste of fighting power through avoidable casualties is not a characteristic normally demonstrated by successful militaries. On the other hand, it seems impossible that a state could win a war with a serious opponent unless prepared to accept the risks involved. Yet, perversely per Clausewitz, inflicting casualties on the adversary is the very foundation to the nature of warfare,[33] driving the adversary to a position where capitulation is more attractive than the continued loss of resource. This represents victory in both the political and military space: casualties rather than financial headlines have become the public manifestation of resource invested in military operations. Death, destruction and injury are a core part of warfare and appear likely to remain so.

Despite this, a notable nobility has developed in political and military narratives regarding casualty aversion beyond the wasted resource and costs associated with repatriation and/or long-term treatment. Yet, it is also clear that a state displaying a high casualty-averse policy or position can undermine its own strategic posturing: if casualties matter more than the campaign, and the adversary understands this and messages accordingly, a state can be forced from the conflict without the requirement for a decisive military engagement (Spain in Iraq in 2003–04 is a fitting example).[34] Such manifestations of policy undermine any

---

[31] Hugh Smith, 'What Costs Will Democracies Bear? A Review of Popular Theories of Casualty Aversion', *Armed Forces and Society* (Vol. 31, No. 4, 2005), pp. 487–512.
[32] See, for example, *fs*, 'Attrition Warfare: When Even Winners Lose', 2017, <https://fs.blog/2017/07/attrition-warfare/>, accessed 31 March 2021; Karl Walling, 'Thucydides on Policy, Strategy, and War Termination', *Naval War College Review* (Vol. 66, No. 4, 2013), p. 24.
[33] See Beatrice Heuser, 'Introduction', in Carl von Clausewitz, *On War*, translated by Michael Howard and Peter Paret (Oxford: Oxford University Press, 2008), p. xxviii.
[34] 'The rich man's option is to sanitise war; the poor man's is to make it even more horrendous than it is'. See Christopher Coker, *Humane Warfare* (London: Routledge, 2001), p. 65.

illusions of a credible deterrence policy, whether with irregular armed groups or nuclear armed challengers.[35]

Such noble considerations by political and military leaders can have significant consequences not only for committing to an intervention, but also in the decision-making around how campaigns are waged. The critique applied to US forces in Bosnia was an inability to interact with the population, remaining inside their well-protected superbases and patrolling in armed convoys. This also became the UK's experience in Iraq (hunkered down in Basra airport and deserting the people of Basra).[36] However, the US model for force employment changed markedly after 2001, when the willingness to accept mission risk and cost in blood and treasure were reversed. These examples point to the nub of the issue: that the public – and thus political – will to fight is fickle. The merits of the campaign in public eyes determine an attitude to casualties that is not fixed, but highly dynamic.

These are worthy considerations. Yet, part of the existing literature and research around Western casualty aversion makes a case that it is not about political repercussions or decisions, but rather about changing public attitudes to the use of force and the perpetration of violence by societies at large. Eyal Ben-Ari's study in 2005 noted evidence of an increasing trend of casualty aversion across Western democracies.[37] Given such findings, others have posed the most pertinent question: can democracies any longer tolerate casualties?[38] For militaries, evolving to political demands is a natural evolution in campaign planning, but the shifting attitudes to casualty-averse policies has accompanying operational implications.

Since 1998, the UK Defence Planning Assumptions have codified part of the risk calculus for operations as an expectation of numbers of dead and injured that will be acceptable to the government.[39] As a driving planning and force design factor, the amount of risk for an operation became expressed as the number of deaths, casualties and losses expected in set scenarios. The MoD's planning assumptions represented the department's

[35] Justin Bronk, 'The Weakness of "People" in Deterrence', *RUSI Commentary*, 18 December 2019.

[36] Ben Barry, *Blood, Metal and Dust: How Victory Turned Into Defeat in Afghanistan and Iraq* (London: Osprey, 2021).

[37] Eyal Ben-Ari, 'Epilogue: A "Good" Military Death', *Armed Forces & Society* (Vol. 31, No. 4, 2005), p. 662.

[38] Smith, 'What Costs Will Democracies Bear?'.

[39] Peter Roberts and Ben Barry, 'Episode 44: Evolving the Western Way of War Into (and Out of) COIN', RUSI Western Way of War podcast, 15 April 2021, <https://rusi.org/podcasts/western-way-of-war/episode-44-evolving-western-way-war-and-out-coin>, accessed 20 October 2021.

own book of scenarios for using operational analysis to test force structures and equipment, in which UK military casualties were deemed to be an important element of the calculus that would determine success. Different levels of fatalities and casualties were assigned for different types of mission and differing scales of effort. Whilst there appeared to be little detailed analysis behind them, the figures were representative of the Falklands conflict in 1982, Northern Ireland during The Troubles and UK Balkan experiences in the 1990s. The belief has been that casualty risk aversion is a critical determinant in political decisions to undertake any specific campaign, and that the military might be expected to guarantee a risk level to deployed forces that bounded missions and the part that UK forces could play in a wider campaign.[40]

There is some evidence to underwrite such claims, and an accompanying set of expectations in infrastructure planning for medical and psychological casualties that saw the UK remarkably unprepared for the losses in Afghanistan and Iraq. Part of the problem with managing fatalities, the wounded and their families from Iraq and Afghanistan was the apparent lack of preparedness of both Selly Oak and the rehabilitation and service welfare organisations to manage the consequences. They gave the impression that they had never envisaged the high level of steady casualties or were unprepared to quickly ramp up, or both.[41]

Considerable efforts have been made to reduce the risk to personnel deployed in war zones, specifically those from the armed forces. In order to minimise casualties, there has been an increasing appetite to employ contractors instead of government employees: an approach as clear in Russia as it is across Western governments.[42] Yet, those measures have not been sufficient in interventions when the deployment of sovereign troops has been required. In such circumstances, and against such perceptions of casualty intolerance, the implications on decision-making, planning and military activities have been significant. Now underpinning the entire approach to the UK's campaign planning, response options, and assumptions of endurance, risk and appetite, this idea of a political and societal obsession with casualties has become inculcated in military life. The result has been a desire that any military operation needs to be not only high reward, but low risk: a dynamic that is almost impossible

---

[40] Cornish, 'Myth and Reality'.

[41] Richard Dannatt, *Leading from the Front* (London: Corgi, 2011), pp. 342–57.

[42] R Kim Cragin and Lachlan MacKenzie, 'Russia's Escalating Use of Private Military Companies in Africa', Institute for National Strategic Studies, 24 November 2020, <https://inss.ndu.edu/Media/News/Article/2425797/russias-escalating-use-of-private-military-companies-in-africa/>, accessed 1 March 2021.

to achieve. In reality, this approach to casualties has put UK military campaign plans into a state that can only deliver low rewards from low risks. It also places the UK, like other Western militaries, in a position where a reliance on technology is required to match the risk of deploying – let alone seek any type of competitive edge or deliver success. This is a conundrum that the UK MoD and successive governments have developed for themselves; it is hard to find examples of successful military operations on the basis of such planning assumptions in the real world.

The narrative and considerable literature about casualty-averse policies in Western democracies feeds a self-fulfilling prophecy about deployments and military interventions, as well as the conduct of military operations. Yet, it is valuable to review the claims made above against the types of interventions in which they have been made. There is a case that in being selective about which conflicts and wars have been examined for evidence to support such presumptions, incorrect deductions have been made about the nature of risk held by political leaders. This is not new. In the US in 2004, Richard A Lacquement Jr wrote a compelling article in which he addressed the question: 'How does casualty sensitivity affect the pursuit of American national security objectives?'[43] He was reflecting on evidence given by Senator John Glenn in 1997 – a different set of military conditions for US deployments, and now cognisant of not just the lessons of Somalia and Haiti, but also of the 9/11 attacks and military operations in Iraq and Afghanistan. There is also evidence of considerable differentiation in views between military personnel, civilian leaders and the general public. The results of casualty appetite from the public might shock but are worthy of note.[44] The idea that the public is more willing to accept greater casualties than either military commanders or the political 'elite' (as defined by the polling) is not just reflective of a US audience; UK evidence is remarkably similar.[45] In a 1999 poll, the differences between opinions were stark and revealed that the public understood a difference between national security imperatives and casualties. Asked about acceptable casualties in preventing Iraq from obtaining WMDs, the military considered 6,016 as

---

[43] Lacquement Jr, 'The Casualty-Aversion Myth'.

[44] See Triangle Institute for Security Studies, 'Project on the Gap Between the Military and Civilian Society: Digest of Findings and Studies', Conference on the Military and Civilian Society, Cantigny Conference Center, 1st Division Museum, 28–29 October 1999.

[45] Johns and Davies, 'Civilian Casualties and Public Support for Military Action'.

realistic, whilst the civilian leadership strata opted for a figure of 19,045. The mass public considered 29,853 military casualties acceptable.[46]

Despite the evidence, arguments continue to be made that UK views and opinions differ considerably from those in the US. Much of this might be attributed to the UK's obsession over deciphering wars of 'choice' versus 'necessity'. There is also an idea that UK society is a fast adaptor, especially in the case of necessity.[47] The public may indeed be able to make nuanced judgements and engage in a discussion about national security, resourcing of defence and the acceptance of casualties *provided* they are able to view the need as having legitimacy. There is therefore a wider implication that the government, rather than running away from public scrutiny, should be more prepared to make the case for the use of force.[48]

The latter becomes important given the British experience of casualties in combat when contrasting the entire Afghanistan campaign (454 soldiers) and the First World War.

> A senior British military colleague argued that the public would never again tolerate a war with significant military casualties. He made the point that the 454 soldiers killed in Afghanistan represented 'a quiet day on the Western Front' during the First World War. Some basic maths suggests that Britain lost an average of 341 men each day on the Western Front between August 1914 and November 1918.[49]

Given the rhetoric about China and Russia in Prime Minister Boris Johnson's speech at the Munich Security Conference in February 2021,[50] the near- to mid-term defence and security challenges for the UK may require greater sacrifice than has been experienced for two generations. Research does not just indicate that the casualty-averse nature of society and its politicians may have changed, but that this would revert to a norm of behaviours and attitudes rather than the aberration that has been the

---

[46] Triangle Institute for Security Studies, 'Project on the Gap Between the Military and Civilian Society'.

[47] Cornish, 'Myth and Reality'.

[48] Hew Strachan and Ruth Harris, *The Utility of Military Force and Public Understanding in Today's Britain* (Santa Monica, CA: RAND, 2020).

[49] Tim Willasey-Wilsey, 'Does the Pandemic Tell Us Anything About War Casualties?', *RUSI Commentary*, 12 January 2021.

[50] HM Government, 'Prime Minister's Speech at the Munich Security Conference: 19 February 2021', 19 February 2021, <https://www.gov.uk/government/speeches/prime-ministers-speech-at-the-munich-security-conference-19-february-2021>, accessed 20 October 2021.

Western outlook since the 1990s.[51] Whilst such a shift would not be immediately evident to military planners or personnel, the implications would be significant.

Conclusion

The proposition of this chapter is that the will to fight multiplies the effect that a state can exert through the use of force. If the sacrifice is considered worthy, it seems that society both accepts the requirement for casualties and a willingness to make sacrifices, but that they also have greater resilience than they are given credit for. Instead of the deeply fragile ecosystems that they are made out to be, many Western states have been tested by systemic and long-term shocks and have proved robust.

It also appears clear that the will to fight in Western states depends greatly on the context, the adversary and the gains that might be made. Society is unlikely to accept the same sacrifices for a humanitarian mission in Mali as it would to defeat a Russian attack on a major city. The other highly dynamic variable appears to be the state of the narrative of such campaigns: an extremely well-articulated and compelling narrative seems to increase a state's will to fight more than an ill-considered and reactive set of lines-to-take.

In understanding these factors, it also needs to be acknowledged that neither is fixed but instead is underpinned by another set of variables, including the national infrastructure, system of alliances, ability to pay for support and resources, changing demographics, and national strategic culture. None of these are immutable, and each relies on variables and dynamic factors.[52] The key is that the state has agency in shaping its populations' will to fight. But if governments lack sufficient confidence in their causes to justify them publicly, it should be questioned whether the cause is indeed worthwhile. The idea that a figure could be placed on casualties in any campaign is foolhardy and undermines the realities of public concern and political national security calculations.

The implications for the military – in terms of undertaking operations, offering options to political leaders, and the expected risk appetite in their missions and tasks that pose a threat to life – are significant. Two key facets of a new dynamic are most important.

---

[51] Mueller, *War, Presidents and Public Opinion*, pp. 60–69; Steven Kullclay Ramsay, 'The Myth of the Reactive Public: American Public Attitudes on Military Fatalities in the Post-Cold War Period', in Everts and Isernia (eds), *Public Opinion and the International Use of Force*, p. 205.

[52] Martin van Creveld, *The Transformation of War* (New York, NY: Free Press, 1991).

First, the philosophy of higher-risk, high-reward missions (which the UK military have a solid history of delivering) once again becomes possible – in the right campaign. Not only is this more likely to deliver campaign success than an approach determined by the currently employed low-risk, low-gain models, but it would also free the UK from the shackles of technological reliance in order to mitigate risk. Alongside this new appetite for higher-risk missions would be the attractiveness that the UK forces would have to its allies, and the niche this would give the UK in any coalition force design. With such a niche, and for those willing to take such risks, comes influence and bragging rights with pre-eminent allies such as the US.

Second, one of the key vulnerabilities of Western societies to coercion by competitors and adversaries disappears. If casualty aversion and societal resilience are no longer dominant factors in determining response options, those actions designed by belligerents to unhinge an opponent's forces and their will to succeed will not be a factor in deciding the campaign outcome. The result is a more resilient force design model, able to weather well-prepared tests against a 'Western way of war' and less susceptible to leverage from external actors.

It would be sensible to assume that adversaries will seek to increase the imposition of casualties on Western states on the understanding that this will be the critical vulnerability that forces capitulation in any engagement the West undertakes. Similarly, the evidence seems to point in the direction of exploiting the multiplier effect of a positive will to fight, and engaging in a longer, more complex relationship with the public on the topic of national security. Simply putting a number on society's willingness to fight and refusing to engage with the dynamics of the variables is unconscionable.

VI. IN SPACE, NO ONE WILL SEE YOU FIGHT

ALEXANDRA STICKINGS

In 2019, the establishment of the US Space Force as an independent military service reignited various long-running debates on the nature of future conflict in space.[1] While specialists in military space policy understood the formation of Space Force as a reorganisation of existing capabilities and missions, other narratives soon emerged in the broader security and defence sphere. Most centred on fears of direct kinetic conflict and increasing weaponisation of assets in orbit; high-powered space lasers which could target enemy satellites, military bases on the Moon, and the spectre of 'space marines' in the shape of armed service personnel routinely being deployed in orbit.[2] Beyond providing a target for political commentators and late-night comedy show hosts, these narratives distract attention from the real and important issues raised by likely confrontation in the space domain during future conflicts. Specifically, it is necessary to counter the idea that kinetic warfare in space will be a central and early feature of future state-on-state wars.

The idea of kinetic conflict in space is not new. It became apparent during the late 1950s that satellites could provide enabling capabilities for terrestrial military operations. Consequently, both the US and the Soviet

---

[1] John J Klein, 'The Creation of a U.S. Space Force: It's Only the End of the Beginning', *War on the Rocks*, 2 January 2020.

[2] See, for example, Nathan Strout, 'The Space Force Wants to Use Directed-Energy Systems for Space Superiority', *C4ISRNet*, 16 June 2021, <https://www.c4isrnet.com/battlefield-tech/space/2021/06/16/the-space-force-wants-to-use-directed-energy-weapons-for-space-superiority/>, accessed 2 August 2021; Ramin Skibba, 'How Trump's "Space Force" Could Set Off a Dangerous Arms Race', *Politico Magazine*, 22 June 2018; Rachel S Cohen, 'Space Force Will Eventually Put Troops in Orbit, Ops Boss Says', *Air Force Magazine*, 29 September 2020.

Union researched, and in some cases deployed, capabilities that could either destroy the other's satellites or deny access to them. During the second half of the Cold War, the use of orbital assets for missile defence was also explored, with the US proposing the Strategic Defense Initiative (SDI), commonly referred to as 'Star Wars'.[3] This original 'space race' took place within the specific context of the Cold War and as such the focus of both major players was on nuclear capabilities. The 1967 Outer Space Treaty banned the placement of WMDs in orbit, and despite the conventional terrestrial standoff and various proxy conflicts, kinetic confrontation did not extend to or break out in orbit.

The US gained the freedom to largely act unchallenged in orbit following the fall of the Soviet Union, and the resulting loss of inertia within the Russian space programme. Over decades, this US supremacy in orbit nurtured notions in the West that space was a 'sanctuary' free from conflict. However, a range of new challengers and rapidly proliferating capabilities means that space can no longer be kept apart from terrestrial discussions on competition and conflict. It is only the character of this conflict, and the ways that it will be realised in space, that remains a question. Understanding the character of different potential forms of conflict in space, as well as the implications of each, can provide some guidance here.

Kinetic Space War and the Question of Sustainability

Anti-satellite (ASAT) missiles are the most obvious example of the militarisation of the space domain, due in part to the ease of classifying them as 'weapons'.[4] ASAT missiles are typically ground-launched, direct-ascent vehicles which physically collide with a satellite in order to destroy it. At present, capabilities in this area are limited to low Earth orbit (LEO), although there is evidence that some states are pursuing the capability to intercept satellites in medium Earth orbit (MEO) and geostationary orbit (GEO).[5]

---

[3] Atomic Heritage Foundation, 'Strategic Defense Initiative (SDI)', 18 July 2018, <https://www.atomicheritage.org/history/strategic-defense-initiative-sdi>, accessed 5 November 2021.

[4] There is no international agreement on what constitutes a weapon in space. It is now commonplace to use the term 'counterspace capability', which covers all threats to space systems, including ground stations and receivers.

[5] Brian Weeden and Victoria Samson (eds), 'Global Counterspace Capabilities: An Open Source Assessment', Secure World Foundation, April 2020, <https://swfound.org/media/206970/swf_counterspace2020_electronic_final.pdf>, accessed 8 March 2021.

Kinetic ASATs were one of the earliest counterpace capabilities to be developed, not least because of the technical crossovers with existing ballistic missile programmes. As with other early space capabilities, the first programmes to examine their utility were run by the US and the Soviet Union, although neither achieved full proficiency in this area. The US successfully tested the first ASAT in the shape of the air-launched ASM-135 in September 1985, but the programme was cancelled the following year.[6] At present, the US does not have a confirmed ASAT programme, but has demonstrated the capability to repurpose SM-3 ballistic missile defence interceptors for this purpose.[7] More recently, China and India have successfully tested this capability, each destroying one of its own satellites in the process.[8] In November 2021, Russia conducted an ASAT test which destroyed one of its satellites in low-earth orbit and caused the crew of the International Space Station to repeatedly take emergency measures when passing through the resulting orbital debris.[9]

Despite these tests, and the evidence that additional states are likely to pursue ASAT capabilities, questions remain as to the likelihood of such an asset being deployed in practice. The nature of the space environment and the ways in which states monitor and regulate their space assets impose limitations on the practical effectiveness of kinetic ASATs. The first of these is the overtly aggressive and escalatory nature of ASAT use. Physically destroying the satellite of an adversary in this way generates a distinctive signature that will ensure near-instantaneous detection and attribution. Such a blatant attack on strategic infrastructure would certainly elicit a response. However, without historical precedent or recognised international conventions to draw on, the nature of that response, and the domain in which it might take place, would be hard to predict. This uncertainty is likely to cause hesitancy in those states that possess kinetic ASAT technology to use it in practice. Any potential kinetic ASAT use must also be considered within the context of a broader conflict or crisis. Such a directly escalatory act would not only be a leap into the unknown but also contradict the likely desire of both belligerents

---

[6] Paul Glenshaw, 'The First Space Ace', *Air and Space*, April 2018.

[7] George Galdorisi, 'U.S. Navy Missile Defense: Operation Burnt Frost', Defense Media Network, 18 May 2013, <https://www.defensemedianetwork.com/stories/u-s-navy-missile-defense-operation-burnt-frost/>, accessed 1 April 2021.

[8] Shirley Kan, 'China's Anti-Satellite Weapons Test', CRS Report for Congress, 23 April 2007, <https://fas.org/sgp/crs/row/RS22652.pdf>, accessed 7 March 2021; Ashley J Tellis, 'India's ASAT Test: An Incomplete Success', Carnegie Endowment for International Peace, 15 April 2019, <https://carnegieendowment.org/2019/04/15/india-s-asat-test-incomplete-success-pub-78884>, accessed 8 March 2021.

[9] *BBC News*, 'Russian Anti-satellite Missile Test Draws Condemnation', 15 November 2021.

to limit the geographical and temporal boundaries of any future state-on-state conflict – at least in its early stages.

The second limitation concerns changes in the ways in which space assets are developed and deployed. Kinetic ASATs can only be used against a single satellite. If that satellite alone is providing a capability to the user, destroying it could provide a significant advantage. However, the architecture of space systems is changing. There has been a continued move towards diversified and disaggregated space systems, where constellations of multiple satellites, at times covering different orbits, are used to provide a specific capability.[10] Therefore, the loss of one satellite does not necessarily mean the loss of the capability, and so the user of the kinetic ASAT must make a cost/benefit analysis to determine the impact that could be made. Of course, destroying a satellite, regardless of what impact it has on the provided capability, has symbolic and potential political utility, but this would also be subject to the first limitation mentioned, that of response and escalation.

The final limitation is that of space debris. Kinetically destroying a satellite causes it to break up into thousands of parts. It is impossible to know how many pieces will be created in such an event, nor their size and the direction in which they are sent. This, coupled with the realities of orbital mechanics, means that there is no way of determining the true extent of collateral damage that might be done to other space assets, including those of the perpetrator.

The creation of orbital debris fuels perhaps the greatest concern surrounding a kinetic conflict in space; the potential triggering of a debris-collision cascade in the form of the Kessler Effect. Such an event would render the specific orbit unusable for many years, and limit or perhaps even prevent actors from reaching other, higher orbits. The 2007 Chinese test, which was carried out at an altitude of 863 km, produced more than 3,000 pieces of observable debris. Many of these are expected to remain in orbit for decades and pose a long-term collision risk to other spacecraft, which could potentially lead to additional debris.[11] In contrast, the Indian test in 2019 was undertaken at a deliberately low altitude of 280 km, which meant that the resulting debris will not last as long.[12] This has led some to

---

[10] Todd Harrison, 'The Future of MILSATCOM', Center for Strategic and Budgetary Assessments, 24 July 2013, <https://csbaonline.org/research/publications/the-future-of-milsatcom/publication/1>, accessed 1 September 2021.

[11] Brian Weeden, '2007 Chinese Anti-Satellite Test Fact Sheet', Secure World Foundation, updated 23 November 2010, <https://swfound.org/media/9550/chinese_asat_fact_sheet_updated_2012.pdf>, accessed 7 March 2021.

[12] Brian Weeden and Victoria Samson, 'India's ASAT Test Is a Wake-Up Call for Norms of Behavior in Space', *SpaceNews*, 8 April 2019, <https://spacenews.com/

label the Indian test as more 'responsible' and perhaps the way in which any future tests should be carried out. It certainly did not elicit the same international condemnation as the Chinese test.[13] However, any ASAT use against an adversary satellite for military purposes in a future conflict would not be able to take this sort of debris-limitation measure since the altitude of the interception would be dictated by the target's altitude. As such, the unintended consequences of ASAT use could also lead to unpredictable further escalation and the broadening of terrestrial hostilities. This is likely to constrain their utility as a practical tool in future conflicts.

In terms of kinetic space conflict, it is also necessary to examine the narrative that armed service personnel will be in orbit; the ubiquitous 'space marines'. It is true that a high percentage of astronauts, from all states, have been either serving or retired military officers. The early US astronauts were drawn from the test pilot cadre, and Russia selected its first cadre of cosmonauts from Soviet air force pilots.[14] However, the missions which they conducted were generally about science and exploration. Today, NASA astronauts who are serving military officers are seconded to the Agency, and as such are not considered to be representing the military during their time in orbit.

Nevertheless, there was some concern recently when NASA astronaut Mike Hopkins, a US Air Force officer, was sworn into the new Space Force whilst on the International Space Station.[15] This blurring of the line between military and civilian activities highlights what many are worried about: that military activities may increasingly be carried out under the pretence of civilian exploration and experimentation. There is a fear that crewed military missions being carried out routinely by competing powers might lead to direct conflict in orbit. However, beyond the damage that such a confrontation could do to the orbital environment and the prospects for long-term sustainability, there are other challenges that suggest caution is required in assessing such narratives.

First, crewed space platform capabilities are still fairly limited. Despite a constant human presence in orbit aboard the International

---

op-ed-indias-asat-test-is-wake-up-call-for-norms-of-behavior-in-space/>, accessed 5 March 2021.

[13] *Hindustan Times*, 'No Global Heat on India's ASAT Missile Test', 29 March 2019.

[14] NASA, '60 Years Ago: Soviets Select Their First Cosmonauts', 25 February 2020, <https://www.nasa.gov/feature/60-years-ago-soviets-select-their-first-cosmonauts>, accessed 6 March 2021.

[15] NASA, 'NASA Astronaut Mike Hopkins Transfers to US Space Force While Aboard International Space Station', 18 December 2020, <https://www.nasa.gov/feature/nasa-astronaut-mike-hopkins-transfers-to-us-space-force-while-aboard-international-space/>, accessed 6 March 2021.

Space Station (ISS) since 2000, human activity is limited to work onboard the station or space walks in its near vicinity.[16] While the ISS does possess some manoeuvring capabilities to avoid debris in its path, there are no propulsion systems that could enable spacecraft to 'fly' around in orbit in the manner which mainstream science fiction has normalised in the minds of many. Manoeuvring and human activities remain at the mercy of orbital mechanics. How, for example, might weapons be developed that do not have recoil and could therefore avoid altering the velocity and movement of a platform or astronaut after firing? While specific technologies that might solve some of the impediments to a 'shooting war' in orbit could potentially be developed, they would have to successfully compete for limited resources with other capabilities that are of more immediate practical use.

Crewed military missions would also likely be seen as crossing a line in terms of 'militarising' or 'weaponising' space. Although there is general acceptance that space is a military domain, the vast majority of activities undertaken are those that support terrestrial operations and are not overtly aggressive towards others. It is this that keeps the balance in space between the major state actors. Introducing a new parameter such as orbiting service personnel could upset this balance, without offering immediately obvious benefits to offset the potential negative repercussions.

It is evident, therefore, that kinetic conflict in space is neither inevitable nor of great military use, and that despite the rhetoric of some of the bigger military space powers, there are strong incentives for most states to avoid it if possible.

Non-Kinetic Capabilities and Sub-Threshold Activity

The many capabilities offered by orbital assets, and the role that space plays in supporting military and national security activity, has led to a reliance on space that is only growing. A kinetic conflict that could lead to the worst-case scenario of a Kessler Effect, affecting satellites covering the whole range of not just military but also commercial and civil activity, would have severe repercussions for all states, affecting everything from Earth observation satellites in LEO to the ability to reach MEO and GEO and replace the navigation and communication satellites that are found in these orbits.

---

[16] Derek Richardson, 'The ISS Marks Two Decades of Continuous Human Presence', *Spaceflight Insider*, 2 November 2020, <https://www.spaceflightinsider.com/missions/iss/iss-2-decades-of-continuous-human-presence/#:~:text=The%20International%20Space%20Station%20has,two%20days%20later%20on%20Nov.>, accessed 7 March 2021.

However, the ever-increasing dependence on space also presents an obvious set of vulnerabilities which states could leverage to gain advantages over an adversary in a military context. Put simply, states have much to gain from disrupting or denying their adversaries' access to space. Interrupting secure communications, 'blinding' satellites to limit ISR capabilities or preventing access to position, and navigation and timing signals could all provide significant advantages. As such, many states have been developing capabilities that can do this without the destruction or overt aggression associated with kinetic ASATs. Effects can be realised without long-term damage to the environment, and in a way that makes attribution difficult.

Laser weapons play a part in this, but in a very different way than that suggested by the more sensational public narratives surrounding 'space lasers'.[17] Ground-based lasers can be used to 'dazzle' the optical sensors of Earth observation satellites, an effect that is temporary and reversible. It is used to deny ISR capability, and the potential use of this technique by the Chinese military is of particular concern to the US.[18] Other capabilities include the potential use of high-powered microwave emissions to interfere with internal components of satellites, as well as the jamming and spoofing of GPS signals and cyber attacks.[19] GPS jamming is already commonplace, particularly by Russia, and there have also been a number of incidents of cyber attacks against space systems.[20]

Another technique which spans both non-kinetic and potentially kinetic activity is close approach and rendezvous, which involves satellites with manoeuvring capabilities that can approach, and in some instances make contact with, other spacecraft. Unlike the other capabilities discussed, this is a technology that is being pursued by the private sector for legitimate commercial activities such as debris removal and on-orbit servicing. Although these are being developed by the commercial sector for non-military purposes, there are good reasons for concern that such rendezvous and proximity operations (RPO) could also have military applications. Early Russian anti-satellite weapons research

---

[17] Andrew Liptak, 'France Wants to Arm Satellites With Guns and Lasers by 2030', *The Verge*, 28 July 2019.

[18] Brian G Chow and Henry Sokolski, 'U.S. Satellites Increasingly Vulnerable to China's Ground-Based Lasers', *SpaceNews*, 10 July 2020, <https://spacenews.com/op-ed-u-s-satellites-increasingly-vulnerable-to-chinas-ground-based-lasers/>, accessed 7 March 2021.

[19] For a full account of the various counterspace capabilities possessed by all actors, see Weeden and Samson (eds), *Global Counterspace Capabilities*.

[20] Alexandra Coultrup, 'GPS Jamming in the Arctic Circle', CSIS Aerospace Security Project, 31 March 2020, <https://aerospace.csis.org/data/gps-jamming-in-the-arctic-circle/> accessed 2 August 2021.

focused on RPO-type vehicle tests,[21] and both Russia and China are thought to have restarted development of this class of vehicle in recent years.

The dual-use nature of RPO capabilities is illustrative of the difficulties in regulating the development of many space technologies with potential military capabilities. It is also another reason why states are likely to pursue them, not necessarily at the expense of kinetic effects, but as more practically usable tools. They provide opportunities to disrupt or deny an adversary's access to space, preventing them from using the data and other benefits provided, while being potentially difficult to attribute and having minimal impact on other spacecraft and the orbital environment.

Increasing Commercial Sector Dominance in Space

Private enterprises will soon not only operate the most satellites, but will also be able to field capabilities which exceed those of many national military satellites across a range of areas including imagery and bandwidth. SpaceX is a clear example of this trend. The company's Starlink mega-constellation, which is intended to provide high-strength internet from orbit, already consists of 1,200 satellites and the intention is to launch tens of thousands more.[22] Similarly, the now well-known ability of SpaceX to land and reuse its first-stage Falcon 9 rocket boosters is well ahead of the capabilities of other launch providers and has dramatically increased launch cadence while also decreasing cost per kilogram of launch.

While SpaceX may at present be leading the pack, others are looking to join in. Amazon Web Services and OneWeb are just two examples of commercial enterprises looking to enter the mega-constellation club.[23] Similarly, numerous private launch providers, including Blue Origin, not only regularly launch both commercial and national security payloads but are also looking to move into the reusable launch market, as well as continually developing new launch capabilities.[24] Large financial resources and the nature of their corporate governance structures allow such companies to take risks that are unavailable to national programmes. One example is the contrast between the many failures

---

[21] Glenshaw, 'The First Space Ace'.

[22] Jeff Foust, 'SpaceX Launches Starlink Satellites and Expands International Service', *SpaceNews*, 11 March 2021, <https://spacenews.com/spacex-launches-starlink-satellites-and-expands-international-service/>, accessed 12 March 2021.

[23] Amazon Web Services, 'AWS for Aerospace and Satellite', <https://aws.amazon.com/government-education/aerospace-and-satellite/>, accessed 28 October 2021; OneWeb, 'Connected as One', <https://oneweb.net/>, accessed 28 October 2021.

[24] See Blue Origin, <https://www.blueorigin.com/>, accessed 28 October 2021.

leading up to SpaceX's first successful landing, and its Starship vehicle, versus the budgetary and time overruns of NASA's Space Launch System.[25] Private companies can develop new capabilities much faster than the public sector. As a result, militaries and governments are likely to become increasingly reliant on the commercial sector to provide certain capabilities, which will give the latter increased leverage over time. This is, of course, the way these developments are playing out for the US and its Western allies. It is to be expected that China and Russia will pursue mega-constellations and reusable launch systems, but in these cases the dividing line between military and civilian capabilities is not likely to be as clear. Nevertheless, the trend is still part of a broader disruption to the traditional field of space exploration and exploitation that will affect how militaries and governments are able to act in the future.

The potential for commercial resource extraction on both the Moon and asteroids is judged by some analysts to be sufficiently valuable to produce the world's first trillionaire in the coming decades.[26] Harvesting resources from asteroids and bringing them back to Earth, perhaps even for on-orbit manufacturing, could drastically alter the economics of space activity. It would also potentially change the balance of power between both states and state and commercial entities. Militaries will need to operate in an environment where their priorities are only a subset of a myriad of competing interests, and legal regimes are still in development, with the possibility of non-state actor involvement. Due to the factors already discussed, the negative implications of kinetic military activity in orbit in such a scenario could have even more detrimental effects on terrestrial systems and economies than at present. This is also a reason for caution when discussing possible military installations on celestial bodies such as the Moon. The Outer Space Treaty prohibits any state from placing a claim on an extra-terrestrial body.[27] However, governments are also likely to face major corporate pressure to refrain from military activities which could impact the ability of private industry to access the resources found on such bodies.

---

[25] Ryan Whitwam, 'Former NASA Head Predicts Commercial Rockets Will Beat SLS', *Extreme Tech*, 11 September 2020, <https://www.extremetech.com/extreme/314905-former-nasa-head-predicts-commercial-rockets-will-beat-sls>, accessed 11 March 2021.

[26] Mike Wehner, 'Asteroid Mining Will Produce the World's First Trillionaire, According to Goldman Sachs', *BGR*, 23 April 2018, <https://bgr.com/2018/04/23/asteroid-mining-trillionaire-goldman-sachs-report/>, accessed 11 March 2021.

[27] UN Office for Outer Space Affairs, 'Treaty on Principles Governing the Activities of States in the Exploration and Use of Outer Space, Including the Moon and Other Celestial Bodies, Article II'.

As commercial space activity continues to gather pace alongside the continued advance of military counterspace capabilities, states must also consider how the rules and regulations concerning space activity will change and the effect this may have on orbital military activity. Most discussions have been taking place through multilateral forums at the UN, such as the Conference on Disarmament and the Disarmament Commission. The Preventing an Arms Race in Outer Space negotiations, and the associated Prevention on the Placement of Weapons in Orbit Treaty,[28] both sponsored by Russia and China, have until recently appeared to be the best hope of this community to reach agreements on limiting certain counterspace capabilities. However, the stalling of the talks and opposition from the US have supported a widespread perception that major progress and agreements are impossible, at least within the current international climate. This has fed a narrative that without agreements in place, states that possess the more destructive capabilities will have free rein to wreak havoc in orbit.

One of the difficulties associated with treaties such as this is the definition of what constitutes a 'weapon' in the context of space. This is particularly relevant when considering the numerous non-kinetic capabilities discussed above, which are often functionally indistinguishable from civilian or commercial capabilities. Members of the disarmament and arms control communities, therefore, have been looking at ways in which to create some international agreement that will at least limit the more dangerous activities. This has led to a focus on behaviours rather than capabilities, and most recently has seen some success through the UK-led initiative through the UN First Committee.[29]

While such initiatives are voluntary, and full international uptake is perhaps wishful thinking, it does point to what may be the way forward, and what therefore may provide some limitation to the more concerning potential future activities. First, such voluntary agreements are only a first step. They can help to build trust and confidence, creating a better forum for dialogue, and it is hoped they will be able to lead to more concrete agreements on the uses of certain capabilities. Second, not engaging with or abiding by these agreements could lead to states being perceived as

---

[28] Louis de Gouyon Matignon, 'Treaty on the Prevention of the Placement of Weapons in Outer Space', *Space Legal Issues*, 8 May 2019, <https://www.spacelegalissues.com/treaty-on-the-prevention-of-the-placement-of-weapons-in-outer-space-the-threat-or-use-of-force-against-outer-space-objects/>, accessed 11 March 2021.

[29] Aidan Liddle, 'Disarmament Blog: Space Resolution Adopted', Foreign, Commonwealth and Development Office, 10 December 2020, <https://blogs.fcdo.gov.uk/aidanliddle/2020/12/10/disarmament-blog-space-resolution-adopted/>, accessed 11 March 2021.

pariahs. As more states come to be actively involved in space activities, and their infrastructures become ever-more reliant on space, the need to work collaboratively to ensure safety, security and sustainability in orbit will become more apparent. Not abiding by agreements, while not illegal, could have significant geopolitical repercussions for those states, especially in the context of a tense terrestrial standoff or flashpoint crisis.

Conclusion

Despite the narratives around the inevitability of kinetic conflict spreading to the space domain in the early stages of any future state-on-state conflict, the reality is likely to be more nuanced. Although there is often a danger in approaching space as separate and unlike other domains, there are ways in which its differences must be recognised. Due to the core physics involved, even limited levels of kinetic conflict in space risk greater repercussions on the environment than similar conflict in other domains. The debris fallout that would result from kinetic ASAT or on-orbit weapons use would potentially affect every spacecraft in that and surrounding orbits, regardless of ownership. Difficulties in attribution and proving intent, stemming from technical limitations in tracking space objects and visualising incidents, add a layer of complexity and uncertainty in the case of less overtly kinetic means, including laser, electromagnetic, cyber and RPO-type attacks. However, the uncertainties over retaliatory dynamics and the unavoidably escalatory nature of attacks on core national capabilities outside the geographical constraints of any particular conflict zone mean that most offensive capabilities are likely to have more value as deterrence tools rather than practical ones.

This is not to say that conflict in space will not occur. Indeed, it is already happening. However, most activity will likely continue to comprise manoeuvring for advantage and temporary interference through jamming, or cyber attacks and other non-kinetic effects. States will continue to find ways that limit the ability of their adversaries to access space and to assure their own access, but in ways that do not threaten the long-term sustainability of orbit or risk uncontrolled escalation. As in many other aspects of future state-on-state conflicts, the mutual interests of all parties to avoid unrestricted warfare is likely to mean that tight political constraints remain on kinetic actions in space, at least during the crucial early stages. As such, orbital conflict should not be seen merely as the effects of ASAT missiles or satellites against other space systems. Instead, it should be examined in terms of the effects which more likely non-kinetic competitive behaviours will have on the ability of these systems to enable operations on the ground. Fighting will happen in space, just not in a way we can easily see.

VII. MORE SENSORS THAN SENSE

JACK WATLING

The US Air Force likes to describe the future of command and control in warfare as analogous to Uber.[1] Such a system promises a drastic improvement in efficiency and cooperation across the force. Suppose, for example, that an infantry platoon needed assistance in engaging enemy armour advancing on their position. They could make the request by reporting the target's position, and this could be made available to all potential shooters in the area. These might comprise an artillery battery, an aircraft en route to a target and an aircraft returning from a strike. One could envisage the artillery battery declining the request because they were tasked with counterbattery duties and did not want to unmask their guns. The first aircraft might also decline because they were already tasked with an important strike mission and needed their munitions for that. The returning aircraft, finding that it had munitions left over from the strike, might accept, and the request would no longer be pushed to other units. Alternatively, if the second aircraft is removed from the equation, a higher commander might be envisaged, with access to the options, determining the trade-off between unmasking the guns, or abandoning the strike mission, based on their broader intent.[2] Without such a system, the infantry platoon would have to call for artillery and air support on separate systems. Since the artillery and aircraft would not coordinate with each other, the infantry may receive no support, or support from both.

---

[1] Rachel S Cohen, 'Want to Understand MDC2? Think About Uber, USAF Official Says', *Air Force Magazine*, 23 September 2019.
[2] Congressional Research Service, 'Joint All-Domain Command and Control (JADC2)', 1 July 2021, <https://fas.org/sgp/crs/natsec/IF11493.pdf>, accessed 26 February 2021.

This 'any-sensor-to-any-shooter' system is premised on the seamless transfer of data between units.[3] It comprises the backbone of a capability supposed to deliver command posts with real-time situational awareness from all available sources, removing the fog of war, and centralising data to allow AI to rapidly generate optimised courses of action.[4] The promise is of a military that operates at higher tempo, with greater efficiency, effectiveness and assurance. In the UK military, this attempt for a defence-wide digital backbone is at the core of future visions of how the force will fight. The aspiration is for multi-domain integration.[5]

There are some contexts within which this vision is partially realised and could be extended. The coordination of air and missile defence around a US Navy task group, for example, sees the network of ships identify incoming threats, assure the status of defensive systems and then prioritise defensive systems across the task group to engage separate targets.[6] This prevents each ship independently selecting the same target, or missing threats aimed at other ships.[7] It is important to appreciate, however, the challenges in creating a type of military Uber, and how the trade-offs necessary to make such a system work lead to quite a different capability to the one often envisaged in depictions of future war. Uber functions because the world is now densely dotted with 3G, 4G and 5G masts, while the overhead constellation of GPS satellites ensures that every device can be constantly tracked.[8] In the context of the naval task group, similar conditions can be met. Its cooperative engagement capability is possible because there is a relatively small number of points in the network, which mainly operate within line of sight of one another. Each ship has a large communications array, and a significant source of power, enabling reliable and rapid transmission of large volumes of data. If the electromagnetic spectrum is contested, high-bandwidth transmission within line of sight can be achieved through free-space

---

[3] Caitlin O'Neill, 'Delivering an On-Demand Sensor to Shooter Warfighting Capability', *US Army*, 29 April 2020, <https://www.army.mil/article/235067/delivering_an_on_demand_sensor_to_shooter_warfighting_capability>, accessed 3 March 2021.

[4] Ministry of Defence (MoD), 'Digital Strategy for Defence: Delivering the Digital Backbone and Unleashing the Power of Defence's Data', April 2021, pp. 14–15.

[5] MoD, 'Joint Concept Note 1/20: Multi-Domain Integration', November 2020.

[6] Johns Hopkins Applied Physics Laboratory, 'The Cooperative Engagement Capability', *Technical Digest* (Vol. 16, No. 4, 1995), pp. 377–96.

[7] For example, as led to the loss of the *SS Atlantic Conveyor* during the Falklands conflict. See Max Hastings and Simon Jenkins, *The Battle for the Falklands* (London: Pan Macmillan, 2012), p. 286.

[8] Open Cellid, 'The World's Largest Open Database of Cell Towers', Unwired Labs, <https://www.opencellid.org/#zoom=6&lat=29.59&lon=-76.03>, accessed 3 March 2021.

optical communication.[9] Consequently, there is a limited ability to disrupt or jam the system. However, those conditions are not universal across domains.

This chapter seeks to outline the different frictions by domain and explain the limits on exploiting an interconnected, all-domain, digital command-and-control (C2) system. While useful, the gains are in many contexts iterative rather than transformative, while dependence on such a system across domains will risk creating numerous single points of failure that can be targeted by adversaries, increasing the vulnerability of the joint force.

Large-scale land operations do not offer conducive conditions to maintaining a robust any-sensor-to-any-shooter network. The land equivalent of a naval task group would be a division. A division comprises several thousand vehicles.[10] These vehicles do not have a large source of power, and most will not be within line of sight of one another. This means that their transmissions will often need to be routed through intermediaries, so that moving data from vehicle A to E will potentially also require vehicles B, C and D to receive and transmit the data. Alternatively, masts, larger antenna and dedicated communications systems can be mounted to transmit beyond line of sight. However, all of these will be susceptible to interference. Interference may arise from terrain, but in the land environment, the closeness of engagements also makes units vulnerable to deployable artillery or standoff jammers,[11] which – unlike shipboard communications – can overpower the receivers on many communications systems. Furthermore, electronic warfare allows transmissions to be detected, but large robust transmission points often need to be set up and cannot be used on the move. Therefore, large-bandwidth data transmission will only be intermittently available as the vehicles with this capability set up, transmit and then manoeuvre again to avoid being destroyed.

Beyond these structural difficulties is one of bandwidth. Modern sensors are generating much higher-fidelity pictures of the operating environment than previous generations. Whereas an artillery spotter would previously have called for fire with a scripted set of sentences of voice transmission, including the target's grid reference, today's recce

---

[9] QinetiQ, 'Free Space Optical Communications', <https://www.qinetiq.com/en/what-we-do/services-and-products/free-space-optical-communications>, accessed 27 July 2021.

[10] 'BAOR Order of Battle: July 1989', <https://www.orbat85.nl/documents/BAOR-July-1989.pdf>, accessed 31 March 2021.

[11] RUSI, 'RUSI LWC 2017 – Session 7', 19 July 2017, 21:54–48:00, <https://www.youtube.com/watch?v=_EcrrD1dBhg>, accessed 8 April 2020.

platforms gather thermal imaging, radar data, acoustic signatures and more. This information must be fused to ensure confirmation of targets at significant range. The volume of data that must be transmitted for a targeting solution has therefore increased dramatically. These sensors are also concentrated on reconnaissance vehicles, which are the most geographically dispersed element of the force. The exact size of data files and bandwidth capacity within military networks is sensitive. Nevertheless, it is evident that the volume of data to be transmitted is accelerating faster than the bandwidth to support its transmission. Within the land domain, therefore, while a commander is increasingly able to interrogate portions of the battlefield in ever-greater detail, accumulating a consistent picture of the operating environment is becoming harder.

Without an as-yet-unforeseeable technological breakthrough, the disparity between data volume and available bandwidth in the land environment is likely to become worse. This is partly because air and naval assets, lacking many of the constraints of land forces, will seek to exploit volumes of data that are not supportable on land platforms. As a result, militaries have experimented with two methods for reducing the volume that must be transmitted. The first is to employ edge processing to sort important from unimportant sensor data at source and to only transmit that which is determined to require the attention of higher echelons.[12] The catch with this system is that the decision as to what is important is made separately by the crews and their supporting software. The data that reaches the centre will therefore be incomplete and, perhaps more importantly, inconsistent. As different parts of the recce screen come to contextually derived independent conclusions as to what is important, the information at the centre will not give consistent information. This will mean that the centre will have a list of targets but will lack much of the vital information that could enable them to prioritise the allocation of shooters, since there will be insufficient information to assess the relative priority of support and advantage. That is not an issue if the force has enough munitions to cover all targets, but will create serious problems in a warfighting context where prioritisation is critical.

The alternative method is for the centre to request discrete returns from the sensor screen. In this system, the centre might, for example, request all information regarding enemy main battle tanks identified by its network of sensors. This system also relies on edge processing. The sensors must interrogate the data they have collected, isolate that which

---

[12] David S Alberts and Richard E Hayes, *Power to the Edge: Command and Control in the Information Age* (Washington, DC: Command and Control Research Program, 2003).

is relevant to the request and transmit only the relevant data. In practice, the centre would request a range of information; not just returns on a single type of vehicle. Nevertheless, the problem with such a system is that it will only provide the centre with what it expects or thinks it needs to know. It will not flag the unexpected, provide indicators of the unforeseen or highlight absences. This exposes higher echelons to the risk of being surprised, which may be exacerbated by the illusion of situational awareness created by the feeds of data returning from the sensor screen.[13]

Historically, brigades have fought battles, divisions have worried about the next battle, and corps the battle after.[14] The different lead times at these echelons has provided time for higher commands to receive and ingest reports from subordinate formations, written by their officers, to give a commander a combination of baseline statistics and the human comments that contextualise them. Each commander has therefore had the ability to have a feel for the battle two echelons below, to balance knowns and unknowns, and use this information to prioritise resources for the battle ahead.

On the modern battlefield, the range of systems and the low force densities anticipated means that all echelons are likely engaged simultaneously.[15] The any-sensor-to-any-shooter concept is an attempt to ensure competitiveness in this context by accelerating the accumulation and interrogation of data. But to accumulate a complete dataset will overload the bearer network. Furthermore, accumulating trend data will leave a command vulnerable to surprise and letting the edge report outliers to the centre will not provide the latter with a consistent dataset to form a basis for analysis. The most realistic workaround is for the system to accumulate trend data based on a high-priority list generated at echelon, and for outliers to be reported via standard reporting methods. This, however, raises two further challenges. First, the trend data will accumulate at a tempo that is out of sync with the reporting of outliers. Higher echelons will therefore either make decisions too quickly to take outliers into account or will have the whole process slowed to the existing operating speed.[16] Second, the usual answer for how the volume

---

[13] It also accentuates vulnerability to systematic deception. See US Army, 'FM 3-13.4: Army Support to Military Deception', February 2019, <https://fas.org/irp/doddir/army/fm3-13-4.pdf>, accessed 30 March 2021.

[14] Jack Watling and Sean MacFarland, 'The Future of the NATO Corps', *RUSI Occasional Papers* (April 2021), p. 20.

[15] Watling and MacFarland, 'The Future of the NATO Corps'.

[16] Nick Reynolds, 'Performing Information Manoeuvre Through Persistent Engagement', *RUSI Occasional Papers* (April 2021), pp. 37–43.

of data is to be interrogated at a competitive speed in command posts is through the application of AI.[17] If the system relies on data transfer and human reporting, however, then either the human reporting will need to be translated into a language ingestible by the AI system – which will take time and require supervision – or the AI will produce solutions based on an incomplete dataset.[18] This may produce solutions that are not just incomplete, but – because AI systems pursue optimal solutions, and therefore often make aggressive trade-off decisions – may in fact be wholly inaccurate in its assessment of the balance in the absence of additional information.[19] Imagine, for instance, an Uber app that had no information about road closures or traffic jams. It may assign tasks to drivers that could not reach the customer within a practical timeframe.

None of this invalidates the usefulness of being able to transfer data between domains and systems. Nor does it mean that AI will not be highly valuable in accelerating the planning of logistical resupply or of coordinating batteries from a fire control headquarters. However, in the land domain, the ability to move data from any sensor to any shooter will in practice function as the transfer of data from some sensors to some shooters, some of the time. The concept will improve the efficiency of C2, but the gains are iterative and not transformative. Furthermore, as the differences between the maritime and ground contexts demonstrates, the utility and constraints on this system will change how it is employed in each domain. While the aspiration to enable maritime, ground and air assets for information sharing is sensible and will provide incidental opportunities to secure advantage, it will be subordinated to how each service intends to fight. The frictions of how systems will be shaped by the approach to fighting in each domain are best outlined by reference to an as-yet-unconsidered domain: the air.

Air power fits curiously into the discussion, as air–land and air–sea integration are the most routine examples of multi-domain engagements. The capacity to draw target data from, or push it to, aircraft is the most immediately comprehensible example of joint activity. Yet, this discussion has arguably been shaped by the relationship between air and

---

[17] Congressional Research Service, 'Artificial Intelligence and National Security', updated 10 November 2020, <https://fas.org/sgp/crs/natsec/R45178.pdf>, accessed 3 March 2021.
[18] A problem already facing cyber threat intelligence needing to ingest non-digital feeds. See Sean Barnum, 'Standardizing Cyber Threat Intelligence Information with the Structured Threat Information eXpression (STIX™)', MITRE, 20 February 2014, <http://www.standardscoordination.org/sites/default/files/docs/STIX_Whitepaper_v1.1.pdf>, accessed 30 March 2021.
[19] RUSI, 'Session Twelve: Innovation and Adaptability', 19 June 2019, <https://www.youtube.com/watch?v=ZYzJIcS36Ls>, accessed 3 March 2021.

land forces that has characterised the post-Cold War period of counterinsurgency in which air forces have been under minimal threat and have therefore been subordinated to the support of ground forces. The cooperative engagement frameworks for ground–air integration are geared towards high-intensity peer-on-peer warfighting. But that context fundamentally changes the priorities for air assets.[20]

In a warfighting context, the threat to air assets will be severe until an effective suppression/destruction of enemy air defences campaign has been completed.[21] The challenge is exacerbated by the shrinking size of air forces, meaning that there is a need for a very low rate of attrition.[22] As it cannot be assumed that adversary ground forces will wait for the air battle to be completed before acting, armies must prepare to fight without available close air support. Even assuming that air assets are increasingly available, however, it must be understood that air power is not likely to prioritise tactical targets. In an environment where air forces bear comparable risk to ground forces, the latter will not be able to dictate the priority of targets. Within air power theory, targets may be divided into the tactical and the strategic.[23] Tactical targets correspond to those in the close and deep battle areas within the land domain because the threat level and impact are comparable from an air force perspective. Strategic targets – such as critical infrastructure, strategic air defence networks, industrial and higher-level C2 facilities – are likely beyond the depth that tactical echelons in ground forces can either detect or strike. This means that while ground forces can expose their positions to send target data to aircraft in a high-intensity warfighting context, it is unlikely that the air force will want to divert assets from sorties against strategic targets to deliver tactical effects.

The sophisticated sensor suites on modern military aircraft are likely to capture a panoply of data that could be highly beneficial to ground forces

---

[20] Justin Bronk, 'Approaching a Fork in the Sky', in Jack Watling (ed.), *Decision Points: Rationalising the Armed Forces of European Medium Powers*, Whitehall Paper 96 (London: Taylor and Francis, 2020), pp. 52–62.

[21] Justin Bronk, 'Modern Russian and Chinese Integrated Air Defence Systems: The Nature of the Threat, Growth Trajectory and Western Options', *RUSI Occasional Papers* (April 2021).

[22] Justin Bronk, *The Future of NATO Airpower: How Are Future Capability Plans Within the Alliance Diverging and How Can Interoperability be Maintained?*, Whitehall Paper 94 (London: Taylor and Francis, 2020).

[23] For example, see Peter R Faber, 'Interwar US Army Aviation and the Air Corps Tactical School: Incubators of American Airpower', in Phillip S Meilinger (ed.), *The Paths of Heaven: The Evolution of Airpower Theory* (New Delhi: Lancer Publications, 2000), p. 215.

as the aircraft penetrate and exfiltrate enemy airspace.[24] However, it must be acknowledged that in data transfer it is the party sending information, not the one receiving it, that bears the risk. While stealth aircraft penetrating enemy airspace will likely detect a range of interesting data points about enemy positions, the period when they are traversing the overlayed threat rings from short-, medium- and strategic-range SAM systems is not the time when they will want to give the enemy an indication of their position. Aircraft such as the F35 can transmit on specialised datalinks without losing their stealth characteristics, but the receiver terminals for the Multifunction Advanced Data Link are highly sensitive and cannot risk being captured through integration into frontline units.[25] These aircraft can therefore pass the relevant data back to the Combined Air Operations Centre, or potentially to an E3 Sentry or E7 Wedgetail. There, the data would need to be translated into a system that could be sent to a receiving ground unit such as the fire-control headquarters at the corps echelon. Alternatively, it could be transferred to an aviation asset such as Wildcat if it were equipped with Link-16.[26] Here, crew could determine what was relevant and pass key points of interest to tactical units on the ground via Bowman,[27] or through Morpheus when it comes into service.[28] Any of these options require two or three translations of the data, as well as it passing through at least four echelons. While this may provide incidentally useful information, it does not constitute the basis for collaborative engagement between air and ground units, since the translation of the data must impose a time lag from capture to delivery.

The situation is made even more problematic if the anticipated trajectory of air operations in great power conflict is considered. The increased range of munitions is steadily pushing large C2 and enabling aircraft orbits further from the battlespace. Even if they are able to stand

---

[24] Douglas Barrie, 'F-35 Situational Awareness: Sensing Isn't Enough', *IISS Military Balance Blog*, 19 March 2019, <https://www.iiss.org/blogs/military-balance/2019/03/f-35-situational-awareness>, accessed 3 March 2021.

[25] Brian W Everstine, 'The F-22 and the F-35 Are Struggling to Talk to Each Other … And to the Rest of USAF', *Air Force Magazine*, 29 January 2018.

[26] Myron Hura et al., *Interoperability: A Continuing Challenge in Coalition Air Operations* (Santa Monica, CA: RAND, 2000), pp. 107–21.

[27] British Army, 'Bowman Radio Update | Director of Information | British Army', 13 May 2019, <https://www.youtube.com/watch?v=AYr4Hix-AL0>, accessed 3 March 2021.

[28] MoD, 'Morpheus Programme: Next Generation Tactical Communication Information Systems For Defence', last updated 18 April 2018, <https://www.gov.uk/guidance/morpheus-project-next-generation-tactical-communication-information-systems-for-defence>, accessed 3 March 2021.

in, such aircraft are easily detectable, and susceptible to electronic attack from ground assets within line of sight that may inhibit their ability to receive data from penetrating ISTAR platforms. As a result of the threat to enabling aircraft, the US in particular is aiming to reduce its dependence on these kinds of platforms. However, this will eliminate a key communications node that at present enables integration with higher-echelon ground formations. If, by contrast, coordination of air assets is pushed back to the Combined Air Operations Centre (CAOC) or Joint Forces Command, this will impose more echelons between the air component and ground units, expanding the lag between detection and receipt of the relevant data.

The situation is markedly different in a maritime context. It should be noted that while much ink has been spilt lamenting the death of aircraft carriers,[29] all navies seeking to project power are currently investing in modernising their carrier capability.[30] Carrier operations will therefore remain at the heart of naval operations for some time. The maritime domain is consequently intrinsically linked to the air domain. This chapter began by considering the cooperative engagement capability within the context of point defence of a naval task group. A carrier strike group would hope to identify threats long before it had to engage in point defence, and to be sensing and striking first.[31] If a nominal combat

---

[29] John Patch, 'Fortress at Sea? The Carrier Invulnerability Myth', *US Naval Institute Proceedings* (January 2010).

[30] See MoD, 'Joint Statement on Carrier Strike Group 2021 Joint Declaration Signing', 19 January 2021, <https://www.gov.uk/government/news/joint-statement-on-carrier-strike-group-2021-joint-declaration-signing–2>, accessed 3 March 2021. On France, see Elysee, 'Notre avenir énergétique et écologique passe par le nucléaire. Déplacement du Président Emmanuel Macron sur le site industriel de Framatome', 8 December 2020, <https://www.elysee.fr/emmanuel-macron/2020/12/08/deplacement-du-president-emmanuel-macron-sur-le-site-industriel-de-framatome>, accessed 3 March 2021. On the US, see Congressional Research Service, 'Navy Ford (CVN-78) Class Aircraft Carrier Program: Background and Issues for Congress', 29 September 2021, <https://fas.org/sgp/crs/weapons/RS20643.pdf>, accessed 3 March 2021. On China, Japan and South Korea, see Felix K Chang, 'Taking Flight: China, Japan and South Korea Get Aircraft Carriers', Foreign Policy Research Institute, 14 January 2021, <https://www.fpri.org/article/2021/01/taking-flight-china-japan-and-south-korea-get-aircraft-carriers/>, accessed 3 March 2021. Russia continues to try, see Paul Goble, 'Moscow's Plans for New Kind of Aircraft Carrier Unlikely to Be Realized', The Jamestown Foundation, 11 March 2021, <https://jamestown.org/program/moscows-plans-for-new-kind-of-aircraft-carrier-unlikely-to-be-realized/> accessed 5 November 2021.

[31] This is the objective of the US Navy's Naval Integrated Fire Control – Counter Air (NIFC-CA) system. See John Andrew Hamilton, 'Navy Conducts First Live Fire NIFC-CA Test with F-35 at White Sands Missile Range', US Army, 29 September 2016,

radius of 400 nautical miles (nm) is assumed for the carrier air wing, alongside a sensor radius from their cruising altitude of a comparable distance, carrier operations sensing targets and potentially engaging them out to 500 nm or further can be envisaged. That range may be extended through aerial refuelling, either using carrier-launched refuelling aircraft[32] or by linking up with air force aerial refuelling aircraft. The former must severely reduce the size of strike packages that can be projected.

The distances and ingrained interdependence of the forces involved have significant implications for the utility of cooperative engagement beyond the defence of the task group. First, the increase in range between different cooperating assets increases the potential lag between action and effect. Aircraft operating at a significant distance from the task group are likely to detect and encounter incoming threats before the surface vessels, and consequently any engagement will have progressed significantly by the time munitions or other capabilities from the task group can engage the target. The coordination of effects over such distances requires careful sequencing, even for missiles such as the SM-3 series which offer hypersonic performance. If it has not been planned to have assets in place to cooperatively engage a target, they may be too far away to do so opportunistically in many circumstances. Munitions are increasingly able to interact cooperatively,[33] as are aircraft in the same strike package, but this technology represents an opportunity for iterative improvements over existing tactics rather than a revolutionary solution to the tyranny of distance in the maritime domain.

The limitations of enablement also reduce the significance of cooperative engagement between the services. For air forces operating over a large maritime area, strike aircraft must either have multiple refuelling points or the strikes must be carried out by long-range bombers. If the former approach is taken, the tankers must be protected, so that to deliver a large strike package against a target requires tiers of enablement at intervals behind. Those tiers of enablers must also remain in place to facilitate the return journey, and so must themselves be escorted. The level of planning, interdependence and precise windows for intersection between aircraft means that these operations are unlikely

---

<https://www.army.mil/article/175940/navy_conducts_first_live_fire_nifc_ca_test_wtih_f_35_at_white_sands_missile_range>, accessed 3 March 2021.

[32] Boeing, 'MQ-25', <https://www.boeing.com/defense/mq25/>, accessed 3 March 2021.

[33] As demonstrated by the US Air Force's Golden Horde programme. See Garrett Reim, 'Golden Horde Collaborative Bombs Hit Four Targets Simultaneously in Test', *Flight Global*, 26 February 2021, <https://www.flightglobal.com/fixed-wing/golden-horde-collaborative-bombs-hit-four-targets-simultaneously-in-test/142655.article>, accessed 3 March 2021.

to be diverted once in progress. They are not going to plan on the basis of potentially diverting assets from their mission to take advantage of opportunities detected by maritime assets. They will pursue their mission, and the data from maritime assets in the operating environment will be used where available to evade or engage pop-up threats on the pre-planned route and for planning subsequent operations. Conversely, data gathered by the strike aircraft may well be passed to maritime forces, but those maritime units will plan operations based on their own requirements rather than changing missions in progress to account for opportunities.

Bombers face a different challenge. Seeking to minimise their signature, they will employ flightpaths carefully designed to penetrate hostile air space and will be avoiding datalink transmissions. Again, they are not likely to divert or cooperatively adjust their missions on the basis of opportunistic information delivered through joint C2 networks. Instead, interservice cooperation in the maritime environment is likely to depend on detailed planning cycles within the air force and navy CAOCs. Constraints are largely imposed by engagement distances and the time required to put enablement in place, and will exceed the length of planning cycles so that accumulating data faster will not necessarily lead to an increased tempo of sorties and strikes.

The limitations to future C2 do not change the fact that being able to transfer data between units from different services is valuable and will improve the effectiveness of the joint force. That units have the ability to share data – even if they do not always choose to – opens up many small but important efficiencies and, by increasing the number of routes by which information can be moved around the battlefield, reduces the vulnerability of communications systems to attacks on critical nodes.[34] The any-sensor-to-any-shooter concept is worth pursuing.

However, the fact that the integration of communications across domains is advantageous does not mean that it will be decisive. As this chapter has sought to explain, the reason why communications architectures for each domain have evolved differently reflects the distinct challenges that forces face and the way in which they fight. These differences in approaches will continue to shape decision-making and ensure that commanders will still need to penetrate the fog of war, especially in the land domain. That any vehicle can theoretically pass information back means that the commander, in determining what information they wish to prioritise on the network, can likely find an

---

[34] Bryan Clark and Timothy A Walton, 'Taking Back the Seas: Transforming the U.S. Surface Fleet for Decision-Centric Warfare', Center for Strategic and Budgetary Assessments, 31 December 2019.

answer to any question about the battlefield that they choose to ask. But being able to pierce the fog of war is different from it being lifted entirely, and as long as bandwidth is limited, there will only be so many questions that commanders can ask.

The problem with the any-sensor-to-any-shooter narrative is that it suggests to commanders that they will soon have a complete picture of the battlefield and be able to access a full set of real-time and assured information. In practice, while it will be increasingly possible to link any sensor to any shooter, this does not mean *all* sensors to *all* shooters, but rather some sensors to some shooters, some of the time. The risk, in short, is that this narrative prevents commanders from accepting the trade-offs intrinsic in what information they prioritise. Unless militaries determine what information is vital, accepting that an incomplete picture will have drawbacks, they risk aspiring to a vision that is unachievable and in doing so miss the genuine advantages that advances in communications technology offer.

CONCLUSION

JUSTIN BRONK AND JACK WATLING

In March 2021, the UK published its most far-reaching reassessment of its foreign and defence policy since 1997. The Integrated Review articulated a vision of the UK as an independent global player, using widespread access to develop economic and political influence, embedded in a strong Western alliance, and at the leading edge of emerging technologies.[1] It stated that Russia was a threat that must be deterred, and that China was a strategic competitor. Whether one agrees or disagrees with its prognosis, the Integrated Review provides a clear articulation of what UK foreign policy aspires to achieve.

The subsequent Defence Command Paper (DCP) was supposed to set out how the Ministry of Defence (MoD) would structure the armed forces to meet the policy demands in the Integrated Review.[2] It failed to do so coherently. This was partly because the forces which can be fielded within the budget available fall far short of what would be required to meet the policy ambition described.[3] However, beyond a lack of fiscal realism, the DCP also demonstrated various conceptual failures which are likely to hamper the ability of Defence to deliver relevant policy options.

One example from the DCP which illustrates some of the real-world consequences of the distortionary narratives highlighted in this Whitehall Paper is the establishment of the British Army's new Ranger Regiment.

---

[1] HM Goverment, *Global Britain in a Competitive Age: The Integrated Review of Security, Defence, Development and Foreign Policy*, CP 403 (London: The Stationery Office, 2021).

[2] Ministry of Defence, *Defence in a Competitive Age*, CP 411 (London: The Stationery Office, 2021).

[3] For more on this trend, see Justin Bronk, 'The Problem at the Heart of UK Defense', *Breaking Defense*, 13 September 2021, <https://breakingdefense.com/2021/09/the-problem-at-the-heart-of-uk-defense/>, accessed 22 October 2021.

The Ranger Regiment was set up as a specialist infantry unit to accompany partners and generate credibility for overseas capacity building, working upstream of conflict to enable local forces to tackle threats at source. There is nothing wrong with this proposition in and of itself. However, the DCP failed to show any self-awareness as to the fact that this strategy had been pursued in many countries throughout the previous two decades of the War on Terror and has failed dismally. The surprise among senior MoD officials when the Afghan military collapsed in weeks during August 2021 underscored the extent of the institutional blindness to the outcomes of this policy. Similarly, extensive training to Yemeni and Saudi forces has failed to produce units that can defeat Houthi insurgents. Training provided to the Malian army has seen the insurgency in Northern Mali expand, and a second insurgency erupt as a partial consequence of Malian military activity, before spreading throughout the tri-border area. There have only been a few isolated successes to offset this widespread failure. The Counterterrorism Service (CTS) in Iraq is a standout example. Afghan Commando units also performed well, though they were unable to compensate for the weakness of the Afghan National Army at higher echelons.

When the principal determinants of success and failure between these various missions are considered, two things stand out. First, the comradeship and shared purpose achieved through accompanying units in combat is vital to success in partner force capacity building, but this is dependent on high-level permissions and risk tolerance. The MoD could have accompanied much more widely across its training missions with or without a Ranger Regiment if it were to accept a higher level of risk to personnel. Conversely, the creation of a Ranger Regiment does nothing to address the cultural risk aversion that has undermined such efforts in the past. If anything, the desire to grant these permissions to a special unit, aligned with Special Operations Forces, can be seen as a way to reduce or even eliminate public discussion of the mission. In this light, the Ranger Regiment appears to be a means of evading the need to proactively shape public perception to support missions and thereby build a national will to fight that would allow for greater risk to be taken. The narrative which holds sway in Defence that the public will not accept casualties leads to military training being hamstrung by inadequate permissions and risk tolerance. These issues are entirely within the gift of the MoD to solve and have nothing to do with budgets. However, any hint of a more mature approach to assessing risk and potential gains is absent from the DCP.

A second failing of training and advisory efforts throughout the War on Terror has been to ignore the principle of concentration of effort. In most training contexts, short-term training teams of around six people were deployed for weeks to deliver courses to large, partnered units. The lessons they delivered rarely stuck. Elsewhere, as with the Afghan National Army, the scale of personnel mobilised dwarfed the number of trainers dedicated to support the institution. In short, however knowledgeable the trainers, what was missing was persistence and mass. By contrast, in Iraq, a critical mass of trainers was achieved for the CTS. The Ranger Regiment offers the potential concentration of specialists to enable the British Army to have more people spending longer with trainees. However, the Integrated Review does little to articulate a hierarchy of priorities. Instead, the DCP exhorts the Army to be more deployed, more of the time, but does not suggest where Defence should place its bets given limited units of action.

The result of these factors is that the success or failure of the DCP as regards its forward presence strategy has much less to do with the name of the unit assigned with the task, or the equipment they carry, than with whether the MoD remains ensnared by its conceptual shortcomings. Success will first and foremost be determined by whether the MoD recognises that there is a critical mass at which partnered operations will succeed and below which they will achieve little or even prove counterproductive. The UK has tried doing 'more with less' in partnered operations, and the results speak for themselves. Second, will the MoD pick a manageable number of operations and justify them publicly in order to develop sufficient public and political will to take the risks that are necessary to ensure the force's credibility? As long as casualty tolerance is assumed to be inherently low, trainers will not be given sufficient boundaries to achieve success on the ground. The Ranger Regiment is an example of why the challenges to distortionary narratives articulated in this paper are necessary; conceptual failings undermine the ability of the new formation to deliver policy-relevant results.

The UK's approach to hostile state actors and strategic competition is similarly weakened by conceptually flawed narratives within Defence. If the UK is to expand its international presence and try to compete with Russia and China for influence or deter Iran from destabilising partners, then it is axiomatic that its military presence abroad will be in tension with the foreign policy interests of these states. If the pursuit of competition short of combat operations comes at the expense of being able to generate credible combat capabilities, however, then there is very little reason for such adversaries to accept a competition space framed to suit UK capabilities. This can already be seen in the extensive use of coercion by Iran against the UK, with tanker seizures, strikes on UK shipping and the sponsoring of harassing fire on bases where its forces are present. Both UK civilians and military personnel have been killed by

Iranian actions.[4] The consequence of this behaviour has been to massively increase force protection requirements, and to fix scarce naval assets to protecting shipping. Similar actions by an increased number of competitors would quickly outstrip the UK's capacity to protect its dispersed forces. In other words, without challenging the narratives on the acceptability of casualties and risk, adversary states could rapidly price the UK out of competing, since without enough assets to provide protection, permissions would rapidly constrain operations.

Distortionary narratives have also fuelled false assumptions within UK Defence that specific actions are escalatory and others are not; some actions are perceived as sub-threshold by nature, while others are axiomatically above the threshold. This misses the fact that the UK has a great deal of agency in dictating where the threshold lies between what it will and will not tolerate. However, the ability to shape the thresholds for various adversary behaviours is in no small part a function of the capacity to escalate with conventional hard power. Limited non-kinetic punishment to impose costs on adversary states can send a clear message as to where the UK believes the threshold should be, but only if backed up by a credible threat of imposing unacceptable costs if the adversary continues to push the threshold. Notably, this still holds even if the cost of escalation is mutually unacceptable. If military actions above a given threshold may be deemed unacceptable to both parties, competition may be conducted sustainably below the threshold. But if one party can be conventionally deterred from responding when an adversary uses force, they will be unable to compete.

At the same time, the false dichotomy between competition and conflict has a tendency in the MoD to relegate warfighting to a space that is apolitical. Because the MoD does not control the political levers of power and struggles to simulate them, warfighting is often discussed as a pure military sphere in which the policy and political constraints that make operating in competition so difficult suddenly dematerialise. This, too, is misleading. It is in no country's interests to deliberately start a third world war. Even immediately before the Second World War, Adolf Hitler had envisaged his campaigns as limited conflicts. He had not expected wide declarations of war over Poland.[5] This is important to

---

[4] Dan Sabbagh, 'UK Servicewoman Killed in Missile Attack on Iraqi Base Is Named', *The Guardian*, 12 March 2020; *BBC News*, 'Man Killed in Tanker Attack Named as Adrian Underwood', 4 August 2021.

[5] Antony Beevor makes a strong case for understanding the Second World War as a series of local and limited conflicts that converged over time, gaining coherence retrospectively. See Antony Beevor, *The Second World War* (London: Weidenfeld & Nicolson, 2012), pp. 1–11.

appreciate because the risk of nuclear escalation and thereby total and mutual defeat ensures that there are very strong incentives for states to politically and geographically limit conflicts that may occur. This means that high-intensity warfighting may well occur in the future without a total war, as in Korea between 1950 and 1953.[6] Such a scenario would entail significant political limitations on what actions can be undertaken even in a state-on-state conflict. US Multi-Domain Operations, for example – envisaging the penetration and disintegration of enemy anti-access and area-denial systems in their own territory – may not be viable under a range of conflict scenarios due to the likely escalation implications. Because of the wider consequences, there are also likely to be severe limitations to what states are prepared to do to one another's space-based infrastructure in the event of war. The potential for high-intensity conflict to occur under geographically and politically limited conditions means that conventional forces able to credibly conduct high-intensity limited conflicts have continued policy relevance even in a world where unrestricted state-on-state warfare remains highly unlikely.

Appreciating that competition and conflict overlap is also important in assessing the potential utility of cyber capabilities. One of the authors has observed instances of military officers assuming that because a cyber capability has been proven against a type of system, it can be called for and delivered as an effect against any system of that type. In reality, cyber capabilities can only function against the specific system in which they have been emplaced. However, this presents a serious challenge to the idea that cyber attacks are 'sub-threshold' capabilities. If they must be embedded in the specific target system to be militarily effective, and the process of embedding the capability has a lead time far longer than the political crisis preceding war, then cyber operations must constitute preparatory activities for war conducted against a state in peacetime. Any cyber capability that would be materially significant to a conflict straddles thresholds between competition and conflict.

Poor understanding of how emerging technologies work, exaggerated expectations of their effects, and – perhaps most importantly – a theory–praxis gap between concepts and what is practically available all present dangers to Defence successfully meeting the requirements in the Integrated Review. It may be argued that the DCP under-resourced current capabilities because the government is betting on future systems. There is undoubtedly a step change in capability on the horizon in many areas of military technology that will reshape the balance of arms across

---

[6] Donald Stoker, *Why America Loses Wars: Limited War and US Strategy from the Korean War to the Present* (Cambridge: Cambridge University Press, 2019), pp. 44–80.

the domains, and in this regard the bet may prove sensible. But it must also be noted that, in wargames for the future force, communications systems tend to work perfectly. Data is always available. Bandwidth management and interference challenges are acknowledged but rarely tested. Meanwhile, units conducting live exercises cannot test capabilities they do not possess. To assume that the resultant systems will work in the field as they have on a map board is dangerous and ignores the huge investment by competing states in systems and operational concepts specifically designed to deny connectivity.

Similarly, the RAF, Royal Navy and Royal Artillery may develop a wide range of complex weapons, and each may be highly effective because of multispectral sensors and swarming capabilities. However, expensive munitions make for smaller affordable stockpiles. In Exercise *Warfighter* in 2021, the British Army fired a simulated 14,000 GMLRS rounds and 45,000 155-mm shells over eight days.[7] This is consistent with historical consumption rates. If ground and maritime forces lack mass and become more dependent on fires, then Defence must have a sufficient volume of them. Thus, at some point, the UK will need to select which high-end capabilities it wishes to prioritise and ensure it has enough launchers and stocks of those capabilities to make a difference. Similarly, a force design premised on seamless coordination under all conditions will have far less combat power in practice (once the adversary and general battlefield chaos cast their votes) than appeared to be available on paper. Without an appreciation of friction in the employment of novel technologies, the force will be overly brittle.

Discussion of defence in the UK has gazed wistfully into the sunlit uplands of future capability for some years now, even as the gap between its ageing and declining fleets of fielded platforms and its aspirations has widened over time. The absence of detail in the description of how future capabilities are to actually work has been consciously and unconsciously exploited to allow hard decisions to be avoided. The UK now faces a critical period where its choices will determine the country's capabilities for the next generation. It is important that those decisions are made with a realistic understanding of the options. Aside from resource constraints, no amount of capability is of any value if it is not properly employed. The authors hope that the conceptual challenges set out in this Whitehall Paper will help ensure that the hard choices confronting UK Defence are approached with clear eyes.

---

[7] Charlie Hewitt, speaking at RUSI's Precision Strike in 21st Century Multi-Domain Operations Conference, 13 May 2021.